U0098434

第一次學做
Cheesecake!
Yummy

前言

跟我學做麵包的女學生，很多人都問：「老師，為甚麼你不教做蛋糕呢？我們很想學做蛋糕啊！」

這便是這本書的緣起了。

學生說，那些外形精緻複雜、味道甜美可口的蛋糕看起來要經過很多複雜的工序，那個領域，大概外行人很難達到吧！我說，麵包也不算很容易，但你現在不是也做得很專業嗎? 可見，事情總是沒有想像中那麼難，只要有決心就可以了。

其實做蛋糕最難的不是造型，而是份量和時間的掌握。和那些可以隨性配搭烹煮的菜餚不同，蛋糕要求材料份量要盡可能準確，時間也要掌控得宜，否則便會影響整體效果。有時做出來的蛋糕，總像欠了點什麼，那可能是糖多了點、或蛋打得不夠久、又或者是麵糰發酵不足等等，以至未能達到最佳效果。

本書會仔細地教大家從基礎做起，初學者宜多做練習，幾次下來，便自有體會，能更純熟地掌握技巧。有了紮實的根基，便可發揮創意、任意造型了，甚至，你也可以多做嘗試，設計出不同新口味的蛋糕來。

書中的蛋糕特地按難度分門別類，讓大家可以一步一步、由淺至深地學習，大家在學習基本知識時，也要多多瞭解材料特性、參考不同配搭的方式，這樣才會得到更多啟發、深入地掌握技巧。

聖誕節又快到了，發揮你的創意，為親友做一個精美蛋糕、和至親度過溫馨時刻吧！

邱勇靈

Preface

Quite a few female students from my bread baking classes asked me: "Sir, why don't you teach us to make cakes? We'd love to learn it!"

It kicked off the production of this cookbook.

They were wondering that it should go through a lot of complicated steps in making delicate and yummy cakes. For the amateurs, it seems hardly make it! I'd say, making bread is not also easy and now haven't you reached professional standard? See, thing are not as difficult as imagined. It is up to your determination.

It is not hard to decorate a cake. How to handle quantity of ingredients and time is pivotal. Unlike recipes of dish in causal style, it requires accuracy in quantity and time in making cakes for desirable results. Occasionally you may find something wrong. It could be due to too much sugar, inadequately beaten egg or under fermented dough.

In this cookbook, you may find the basics in details. Beginners should practice more and will grasp the skills soon. Based on the firm foundation, you may create whatever you like and even, after attempts, design unique, innovative cakes in your very own style.

I have divided the cookbook in sections according to the level of difficulty. You may learn step by step from the basics. And you can understand the features of ingredients and refer to various combinations. So you will acquire the hands-on techniques!

Christmas is around the corner. Be creative. Make a delicate cake for family and friends to share the wonderful moments!

Yung Ling YAU

目錄 Contents

我的第一個起士蛋糕 My First Cheesecake

低難度

中難度

高難度

起士蛋糕房訓練室 Cheesecake Baking Workshop

真假起士事典簿

Cheese Bible

在烘焙甜點時，糕餅師愛用軟起士做甜點和蛋糕，因為軟起士容易與物料和蛋糕糊融合，口感輕盈，質感柔軟細緻，味道清淡卻有獨特風味，又能與許多烘焙材料混合，誘發鮮果香味和平衡味道，為了讓大家多點瞭解起士特質，以配合蛋糕和甜點需求，現在先介紹書中常用的起士：

When making cakes or dessert, pastry chefs love to use soft cheese as, with its smooth and light consistency, it is easy to incorporate and create unique freshness. Soft cheese is also a choice to mix with some other ingredients, to enhance aroma of fruits and keep tastes in harmony. In order to let you know more about the features of cheese and how to match, different types of cheese I use in this cookbook are listed:

1. 奶油乳酪

類別： 美國奶油乳酪，分有低脂、半脂和全脂也有特色風味。

成份： 全脂牛奶或鮮奶油，乳脂含量最少至33%，水分含量不得超過55%。

特質： 溫和淡味、如天般的光滑柔軟質感，含清新檸檬香味。

用途： 可用作沾醬、鹹甜美食、免烤起士蛋糕。

備註： 它源自1872年，國乳品商在重製法式起士時，無意發現，到了1912年James L. Kraft發揚光大，成為舉世知名的費城奶油乳酪。隨著市場需求，風味和品種多樣化，但其中的低脂奶油乳酪因添加了乳清粉，所以會多現其質感含有微粒，不夠滑順。

1. Cream cheese

Type:American cream cheese with different content of fat--low fat, 50% less fat, regular.

Content: Whole milk or cream, with minimum content of 33% milk and 55% water or less.

Consistency:mild, light, glossy and smooth as velvet with fresh lemon flavour.

Use:Dip, sweet or savory dish, chilled cheesecake.

Note: It was discovered accidentally by an American manufacturer in the process of making French style cheese in 1872. In 1912, James L. Kraft modified it and it becomes the famous Philadelphia Cream Cheese. As upon demands from market, more various flavours and types were produced. As mixed with whey powder, low fat cream cheese contains crumbs and is not smooth enough.

2. 馬斯卡邦起士

類別： 義大利新鮮起士。

成份： 它從製造帕瑪森起士的牛奶浮層，即浮出來的鮮奶油作用料。

特質： 豐厚、滑順，油脂含量達75%卻味道柔和。

用途： 義大利甜品，鹹甜美食，它可當作是動物性鮮奶油(double / heavy cream)使用。

備註： 源出於塔斯卡尼(tuscandy)和倫巴弟(lombardy)的馬斯卡邦起士，雖稱為義大利新鮮起士，但嚴格來說它卻不是真起士，只可視作成熟的鮮奶油。因為其質感濃厚和滑順如奶油狀，與奶油乳酪相似，所以仍可說是凝乳起士。

2. Mascarpone cheese

Type:Italian fresh cheese.

Content:Top cream layer of parmesan cheese.

Consistency: Rich, smooth, mild with 75% fat.

Use: Italian style desert, savory and sweet dish,used as double/heavy cream.

Note: It originates from Tuscany and Lombard. Strictly speaking, it is not real cheese. It is mature cream. Its consistency is as rich and smooth as cream cheese.

3. 茅屋起士

類別： 美國低脂起士。

成份： 牛奶。

特質： 柔軟、淡而無味，質感濕潤。

用途： 可烹煮，但常運用於製作小吃和沙拉。

備註： 源產於歐洲，之後傳入美國。它是牛奶的自然凝結物，然後切成小塊，再與乳清一起加熱直至所需質感，瀝去水份，可自由加入鹽、少許牛奶或鮮奶油添加不同味道，變化產品的風味。

3. Cottage cheese

Type: American low fat cheese.

Content: Milk.

Consistency: Soft, plain and moist.

Use: Cooking, especially for making snack and salad.

Note: It originates from Europe and then became popular in America. It is natural milk solid, and then chopped and heated with whey. Drain. Add salt, a small amount of milk or cream to vary tastes.

4. 馬自瑞拉起士

類別：義大利新鮮起士、包裝和輕微煙燻。

成份：水牛奶或乳牛奶。

特質：色澤白晰，柔軟、有 彈力、可伸縮、味道輕盈可拉出細絲狀。

用途：不會添加食物風味，只用作增強食物的彈性和輕柔質感。

備註：傳統的馬自瑞拉起士用水牛奶製造，現在會用乳牛奶製造，兩者的味道都不濃烈，但是乳牛製造起士，柔軟度和細緻質感始終不及水牛奶好。一般情況，它會浸在乳清，保持柔軟鬆軟的質感，要是略煙燻的過程，口感會變得乾而有彈性，至於存放在塑膠袋售賣裝就比較硬和有彈性，只能用作製造薄餅，它的味道不及浸水的起士好風味。

4. Mozzarella cheese

Type: Italian fresh chees, wrapped or lightly smoked.

Content: Buffalo milk or dairy milk.

Consistency: White in colour, soft, bouncy, elastic, stretchy, light in taste, split into fine strips.

Use: Tasteless, only enhancing the elasticity and softness of food.

Note: Traditionally, mozzarella cheese is made from buffalo milk. Dairy milk is commonly used nowadays and it is less tender and smooth. Both types are plain in taste, though. Usually it is soaked in whey to keep its soft consistency. If smoked, it is slightly dry and more elastic. If it is wrapped in plastic, it is only used for making pizza because the texture is harder and more elastic.

5. 乳酪 / 優格

類別：歐洲和美國乳酪，原味、水果味、低脂乳酪，乳製品。

成份：凝結牛奶、酸、乳酸菌。

特質：新鮮、柔軟、味道強烈，有微酸味道。

用途：沾醬、甜品和調味醬。

備註：傳說乳酪是巴爾幹島遊牧民族在數千年前發明，時至今日就在特定環境控制下，由實驗室所調配出的菌種加入牛奶製成。它含有豐富的鈣質、磷和維他命B，不同品種的乳酪的營養價值各異。極低脂乳酪每100克含0.5克脂肪；每100克全脂乳酪含4克脂肪。

5. Yoghurt

Type: European and American yoghurt; plain, fruit flavour, low fat; dairy product.

Content: Solidified milk, acid, lacto bacteria.

Consistency: Fresh, soft, strong flavour, slightly sour.

Use: Dip, dessert or seasoning sauce.

Note: It is said that yoghurt was created by nomadic tribes in Balkan. Now it is produced by mixing milk and lacto bacteria in well-managed environment. It contains rich calcium, phosphorus and vitamin B. Low fat yogurt contacts 0.5g fat/100g while regular yogurt 4g fat/100g.

6. 豆腐

類別：豆製品，硬/板豆腐、軟豆腐。

成份：大豆、石膏粉 / 鹽滷、水分。

特質：柔軟、含豐富豆香，色澤微黃。

用途：可烹煮，鹹甜皆宜。

備註：大豆含 15% 可溶解碳水化合物 soluble carbohydrates、18% 油 oil（其中含0.5%卵磷脂 lecithin）、14% 濕度 moisture、灰質 ash, 其他 other、15% 非溶解碳水化合物 insoluble carbohydrates 和38% 蛋白質 protein。

6. Tofu

Type: Soy bean food, firm tofu, soft tofu.

Content: Soy bean, gypsum powder/brine, water.

Consistency: Soft, rich aroma of soy bean, slightly yellowish.

Use: Cooking savory or sweet dishes.

Note: Soy bean contains 15% soluble carbohydrates, 18% oil (0.5% lecithin), 14% moisture, ash, and other, 15% insoluble carbohydrates, and 38% protein.

我的第一個起士蛋糕
My First Cheesecake

在教學生涯裏，接觸到不同階層的人，有的是業界朋友；有的是有志開業的年輕才俊；有的是白領儷人；有的是家庭主婦，他們為了許多不同原因報讀烘焙課程，問起為何喜歡做糕餅，答案很單純，自己做的糕餅用料足，衛生又好玩，自用或送禮，皆是聊表心意的禮物。

回想我從學徒升到師傅位置，第一個親手做的起士蛋糕的模樣已記不清，只知道從情緒激動而轉為平淡，有點兒麻木不仁的感覺，很無聊啊！雖然如此，我卻因為學員手捧著自製起士蛋糕的喜悅之情，滔滔不絕，興奮又感動地與我分享成品的感受，無形中也觸動我的心，為他們雀躍，重拾當年客人購買我的蛋糕的感覺。

起士蛋糕，你的魅力何在？為甚麼能令女士們如癡如醉，瘋狂地愛上「妳」的滋味？不如，我就把這些令人抗拒不了的起士誘惑，如數家珍地呈現眼前，並按不同難度分為三級制，無論任何人都可因應自己的熟練度而挑上喜愛的款式試做，大家一起跟著做吧！

During teaching years, I met people from different backgrounds: those worked in the same industry, ambitious young adults, office ladies, housewives, etc. They came to pastry class for various reasons. When they were asked why they would like to learn making cakes, their answers were simple---sufficient ingredients, hygienic, enjoyment, gifts for family and friends.

I can't remember how my first cheesecake looked like when I moved on from apprenticeship to a pastry chef. But I was less emotional. My feeling was somehow getting numb. So bored! However, my heart was touched when the students were delighted with their DIY cheesecakes, kept talking with enthusiasm and shared their passion with me. I also got excited and recalled the happy memory that customers bought my cakes years ago.

Cheesecake, why are you so charming? How come all the ladies are crazy about you? Let me introduce this irresistible temptation to the world by dividing recipes in 3 difficulty levels. You may pick what you like. Come do it!

註 Note ：

1. 這本書的起士蛋糕以直徑8吋餅模為標準。
   8-inch cake mould is used in this cookbook.
2. 低難度針對初學者或略有焙烘經驗的人士，以最基本、簡單易操作者為主。
   Level 1: basic and simple procedures for absolute beginners or those have a bit experience.
3. 中難度針對已能掌握基本技巧，進一步把基礎起上蛋糕變化，並學習　些味道層次複雜的製品，挑戰自己。
   Level 2: Some variations and complicated flavour to challenge those who acquired the basics.
4. 高難度針對可以自己處理免烤、烘焙和塔皮類，無論造型和味道趨向商業或專業化，製品更多樣化。
   Level 3: Chilled or baked cake, pie and tart in variety for.

材料 Ingredients

餅底

清蛋糕{ 2片 }
(參閱第121頁)

餡料

奶油乳酪{ 125克 }
糖{ 20克 }
蛋黃{ 2個 }
玫瑰花茶{ 80克 }
牛奶{ 60克 }
百花蜜{ 30克 }
鮮奶油{ 250克 }
吉利丁粉{ 12克 }
滾水{ 70克 }

玫瑰花茶果凍

玫瑰花茶{ 150克 }
吉利丁粉{ 8克 }
糖{ 20克 }

Base

(2) slices base cake
(Refer to p.121)

Filling

(125g) cream cheese
(20g) sugar
(2) egg yolks
(80g) rose tea
(60g) milk
(30g) multiflora honey
(250g) whipping cream
(12g) gelatin
(70g) boiling water

Rose tea jelly

(150g) rose tea
(8g) gelatin
(20g) sugar

難度●●●低

Rose Honey Cheesecake

玫瑰花蜜起士蛋糕

做法 Directions

餡料

1. 將奶油乳酪置於室溫下放軟，然後攪拌至滑順。
2. 吉利丁粉和滾水同置一鍋中，隔水加熱攪拌至完全溶解。
3. 另把糖和蛋黃置於器皿中，隔水加熱拌至滑順，加入玫瑰花茶、牛奶、蜜糖和已溶化的吉利丁粉攪拌均勻。
4. 再與奶油乳酪混合攪拌至滑順融合。

玫瑰花茶果凍

將所有材料拌勻，在熱水上隔水加熱拌至完全溶解。

組合

1. 把鮮奶油打至硬性發泡，加入已拌勻的起士糊。
2. 準備一蛋糕模，先放一片清蛋糕墊於底層，倒入起士糊，放在冰箱中冰2小時至凝固後取出。
3. 倒入玫瑰花茶果凍液於蛋糕上，放冰箱冰30分鐘至凝固，從冰箱取出裝飾即可。

Filling

1. Let cream cheese be softened at room temperature. Blend until smooth.
2. Mix gelatin and boiling water in a saucepan and warm up in hot water. Blend until dissolved.
3. Mix and sugar egg yolk in a container and warm up in hot water. Blend until smooth. Add rose tea, milk, honey and gelatin solution. Mix well.
4. Mix with cream cheese and blend until smooth and combined.

Rose tea jelly

Mix all the ingredients of filling in a container and warm up in hot water. Blend until dissolved and it forms to filling.

Assembly

1. Whip cream cheese until it forms soft peak. Fold in cream cheese mixture.
2. Put a base cake slice in a cake mould. Pour in cream cheese mixture. Refrigerate for 2 hours until it is firm. Take out.
3. Pour in rose tea jelly over cheesecake. Refrigerate for another 30 minutes until it is firm. Take out and decorate. Done.

Q：我見市面上的乾玫瑰花有多種，哪一種比較適合？
I saw a variety of dried roses in market. Which one is best for this recipe?

A：我沒有指定用哪一種，不過台灣品種和法國品種屬天然有機貨，色澤自然，品質有保證，味道比較清香。
It is up to you. Along with fresh fragrance, those from Taiwan and France are organic and come in natural colour. I would recommend them.

Q：荔枝汁應該用罐裝或是用新鮮荔枝自製好呢？

Should I use canned lychee juice or squeeze fresh lychees?

A：新鮮荔枝當季時，又甜又香又實惠，沒有新鮮水果也可用罐裝，味道比較淡和質感較熟一點。至於台灣或澳洲的荔枝就不及中國的荔枝香！台式飲料店就有荔枝果汁，也可使用。

Fresh lychees are sweet and aromatic. Price is also reasonable. You may use canned juice instead, which is lighter and flesh is softer. Lychees imported from China are much better those from Taiwan or Australia. Lychee juice is also available in Taiwanese style bubble tea shops.

難度 ●●● 低

Lychee Yoghurt Cheesecake
荔枝優格起士蛋糕

材料 Ingredients

餅底

清蛋糕{2片}
(參閱第121頁)

餡料

荔枝汁{150克}
荔枝優格{100克}
馬斯卡邦起士{125克}
鮮奶油{250克}
吉利丁粉{12克}
滾水{60克}
糖{20克}

荔枝果凍

荔枝汁{200克}
糖{20克}
吉利丁粉{5克}

Base

(2)slices base cake
(Refer to p.121)

Filling

(150g) lychee juice
(100g) lychee yoghurt
(125g) mascarpone cheese
(250g) whipping cream
(12g) gelatin
(60g) boiling water
(20g) sugar

Lychee jelly

(200g) lychee juice
(20g) sugar
(5g) gelatin

做法 Directions

1. 吉利丁粉、糖和滾水一起置器皿中，隔水加熱拌至溶解。
2. 馬斯卡邦起士、荔枝優格和荔枝汁同置鋼盆中，拌至滑順。
3. 鮮奶油打至硬性發泡。
4. 把已溶化的吉利丁粉與起士糊混合，慢慢拌入已打發的鮮奶油融合，倒入1/2份於蛋糕上後，再放蛋糕片，然後抹平，最後放入冰箱冰約2小時。
5. 將荔枝果凍的材料混合，然後取出蛋糕，淋上荔枝果凍液裝飾，放回冰箱冰至凝固即可。

1. Mix gelatin, sugar and boiling water in a container and warm up in hot water. Blend until dissolved.
2. In a mixing bowl, blend mascarpone cheese with lychee yoghurt and lychee juice until smooth.
3. Whip cream until it forms soft peak.
4. Mix gelatin solution with lychee cheese mixture. Slowly fold in whipped cream until combined. Pour in 1/2 of cream cheese mixture over a base cake slice. Place another base cake slice on top. Level the surface of the cake. Refrigerate for 2 hours.
5. Combine lychee jelly ingredients. Remove the cake from the fridge. Pour in lychee juice for decoration. Refrigerate until it is firm. Serve.

Q：用壓碎的餅乾做餅底，應該用茶餅還是消化餅好呢？
Should I use tea biscuits or digestive biscuits to prepare the base?

A：早期的起士蛋糕會用消化餅做餅底，因為其質感油潤又鬆脆，味道甜中有微鹹，但隨着健康潮流興起，有些烘焙師會轉而使用茶餅或美國Oreo巧可力餅取代；因為它的質感乾爽酥脆，效果也不錯，沒有那麼膩口。
In the early days, digestive biscuits are used for base because it is crisp, moist and sweet with bit saltiness. As healthy diet led the trend, chef started to use tea biscuits or Oreo as they are less oily.

難度●●●低

Lemon Cheesecake

檸檬起士蛋糕

材料 Ingredients

餅底
- 壓碎的瑪莉餅{150克}
- 奶油{80克}

餡料
- 奶油乳酪{250克}
- 糖{20克}
- 蛋黃{2個}
- 動物性鮮奶油{125克}
- 植物性鮮奶油{125克}
- 吉利丁粉{10克}
- 滾水{70克}
- 牛奶{120克}
- 檸檬{半個}

Base
- (150g) marie biscuit crumbs
- (80g) butter

Filling
- (250g) cream cheese
- (20g) sugar
- (2) egg yolks
- (125g) non-dairy cream
- (125g) whipping cream
- (10g) gelatin
- (70g) boiling water
- (120g) milk
- (1/2) lemon

做法 Directions

餡料

1. 將瑪莉餅壓碎後和奶油混合，攪拌均勻，放進8吋餅模裏，抹平。
2. 將檸檬刨皮、榨汁，並將檸檬皮和汁留用。
3. 將奶油乳酪攪拌滑順；糖和蛋黃打發，直至奶白色；再用熱水將吉利丁粉溶解。
4. 煮滾牛奶，加入糖和蛋黃，混合後，再加入奶油乳酪攪拌滑順，最後加入已溶化的吉利丁粉，攪拌均勻。
5. 將植物性鮮奶油及動物性鮮奶油打發，加入吉利丁蛋奶混合物，慢慢攪拌均勻。
6. 加入檸檬皮及汁，快速攪拌均勻，倒入餅模裏，放入冰箱存放約2小時後再取出，裝飾蛋糕。

Filling

1. Combine marie biscuit crumbs and butter. Mix well. Put in an 8-inch cake mould. Press to level the surfacc.
2. Peel and squeeze lemon. Keep lemon zest and juice.
3. Blend cream cheese until smooth; beat sugar and egg yolk until the colour turns to creamy white. Dissolve gelatin in hot water.
4. Bring milk to a boil. Add sugar and egg yolk. Combine. Add cream cheese. Blend until smooth. Finally add gelatin solution. Mix well.
5. Beat non-dairy cream and whipping cream until it forms soft peak. Add gelatin and milk mixture. Slowly blend.
6. Add lemon zest and juice. Blend quickly. Pour in a cake mould. Refrigerate for about 2 hours. Take out. Decorate the cake.

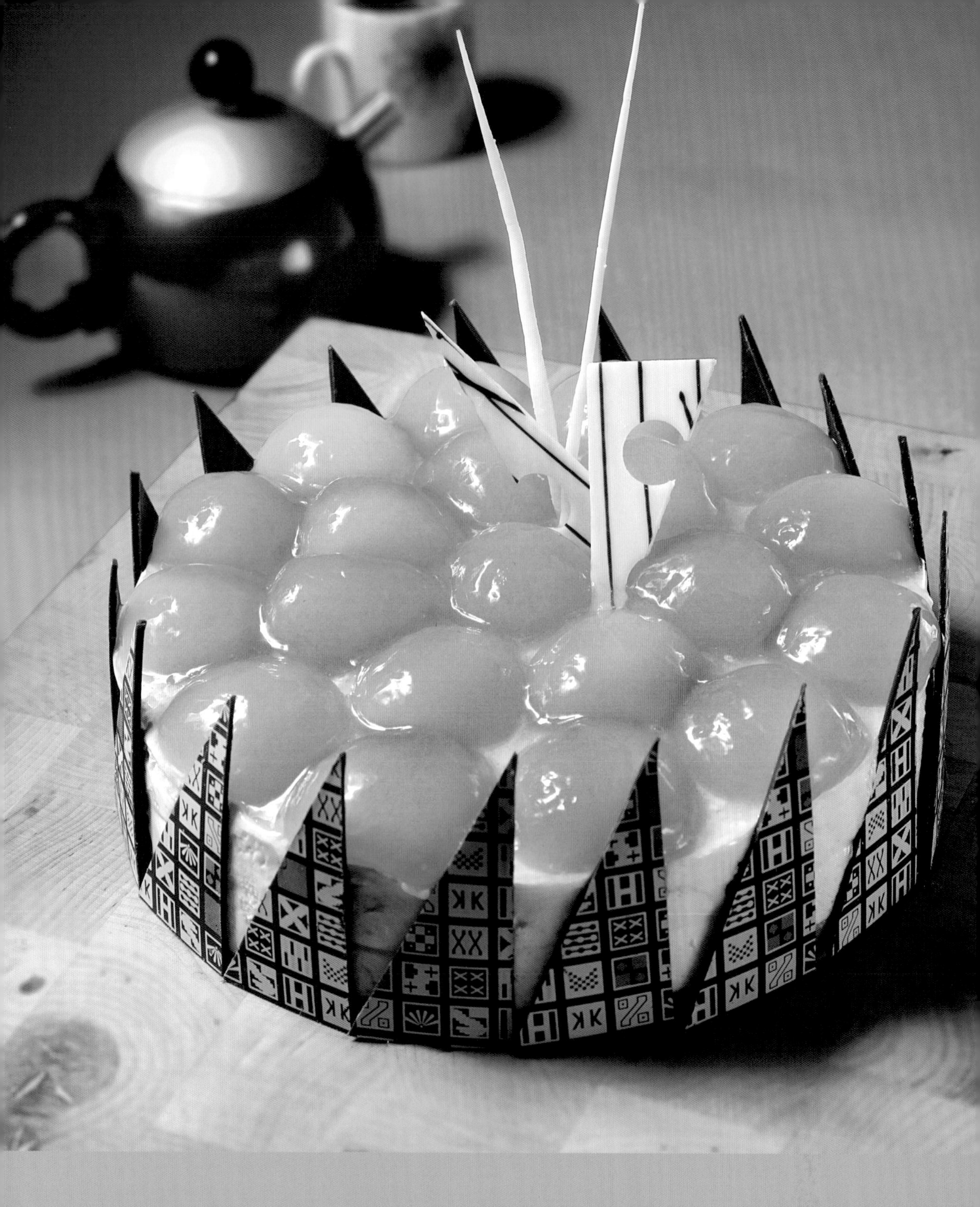

Q：奶油乳酪源自美國，但現在有澳洲出產的奶油乳酪，哪一種的比較好呢？

There are cream cheese products which imported from the United States and Australia. Which one is better?

A：奶油乳酪由美國最先出產，起士味道也頗濃，質感細緻且比較結實；澳洲的奶油乳酪，味道比美國淡一點，質感柔軟，沒美國出產的那麼結實，兩者都很好，只是美國有250克的小包裝，而澳洲就以2千公克為單位來包裝。

Cream cheese originated from the United States. Their products are rich, smooth and firm, coming in 250g package. Those from Australia are lighter in taste and more softened. The size can be as large as 2kg. Both are good options.

難度●●●低

Apricot Cheesecake
黃梅起士蛋糕

材料 Ingredients

餅底
- 壓碎的瑪莉餅{150克}
- 奶油{80克})

餡料
- 奶油乳酪{250克}
- 糖{40克}
- 蛋黃{2個}
- 植物性鮮奶油{125克}
- 動物性鮮奶油{125克}
- 吉利丁粉{10克}
- 滾水{20克
- 牛奶{100克}
- 黃梅果醬{50克}

Base
- (150g) marie biscuit crumbs
- (80g) butter

Filling
- (250g) cream cheese
- (40g) sugar
- (2) egg yolks
- (125g) non-dairy cream
- (125g) whipping cream
- (10g) gelatin
- (20g) boiling water
- (100g) milk
- (50g) apricot jam

做法 Directions

1. 將壓碎的瑪莉餅和奶油混合，攪拌均勻，放進8吋餅模裏，抹平。
2. 將奶油乳酪攪拌滑順；糖和蛋黃打發，直至奶白色；用熱水將吉利丁粉溶解。
3. 煮滾牛奶，加入糖和蛋黃混合後，再加入奶油乳酪攪拌滑順，最後加入已溶化的吉利丁粉，攪拌均勻。
4. 將植物性鮮奶油及動物性鮮奶油打發，加入吉利丁蛋奶混合物，攪拌均勻。
5. 將餡料倒在餅底上，放入冰箱存放約2小時，取出後再用黃梅果醬裝飾餅面。

1. Combine marie biscuit crumbs and butter. Mix well. Put in an 8-inch cake mould. Press to level the surface.
2. Blend cream cheese until smooth; beat sugar and egg yolk until the colour turns to creamy white. Dissolve gelatin in hot water.
3. Bring milk to a boil. Add sugar and egg yolk. Combine. Add cream cheese. Blend until smooth. Finally add gelatin solution. Mix well.
4. Beat non-dairy cream and whipping cream until it forms soft peak. Add gelatin and milk mixture. Slowly blend.
5. Pour in filling over base. Refrigerate for about 2 hours. Take out. Decorate the surface with apricot jam.

材料 Ingredients

餅底

雞蛋{ 2個 }
糖{ 30克 }
麵粉{ 30克 }
酥油{ 10克 }

餡料

馬斯卡邦起士{ 200克 }
糖{ 80克 }
蛋黃{ 2個 }
動物性鮮奶油{ 300克 }
熟香蕉{ 1根 }
吉利丁粉{ 10克 }
滾水{ 70克 }

Base

(2) eggs
(30g) sugar
(30g) flour
(10g) shortening

Filling

(200g) mascarpone cheese
(80g) sugar
(2) egg yolks
(300g) whipping cream
(1) ripe banana
(10g) gelatin
(70g) boiling water

難度●●●低

Banana Cheesecake

香蕉起士蛋糕

做法 Directions

餅底

1. 將蛋和糖打至濃稠；將麵粉過篩後，慢慢加入蛋糊中，攪拌均勻。
2. 加入酥油，攪拌均勻，倒進8吋餅模裏，放入烤箱，用180℃烤10分鐘。
3. 取出蛋糕，放涼，切成2片，放在旁邊待用。

餡料

1. 用滾水將吉利丁粉溶解，備用。
2. 將糖和蛋黃打至乳白色，先加入起士攪拌滑順後，再加入香蕉攪拌滑順。
3. 加入已溶化的吉利丁粉，攪拌均勻；最後加入鮮奶油，攪拌均勻。

組合

將餅底放入餅模中，再加入餡料，放入冰箱冰2小時使蛋糕凝固後取出，用香蕉裝飾。

Base

1. Beat egg and sugar until thickened; sift flour and add to beaten egg. Mix well.
2. Add shortening. Stir well. Pour in an 8-inch cake mould. Bake in an oven at 180°C for 10 minutes.
3. Remove base cake from the oven. Leave to cool. Cut in 2 slices; set aside.

Filling

1. Dissolve gelatin in boiling water; set aside.
2. Beat sugar and egg yolk until the colour turns to creamy white. Add cheese. Blend until smooth. Add banana. Blend until smooth.
3. Add gelatin solution. Stir well. Finally add cream. Combine.

Assembly

Put base cake in a cake mould. Pour in filling. Refrigerate for 2 hours until it is firm. Decorate with banana.

Q：為什麼用雞蛋與糖一起打發的方法做蛋糕，質感有什麼特別的地方？
Why do you beat egg and sugar together? How is the texture like?

A：在烘焙界稱這做蛋糕方法為全蛋法，蛋糕的質感較濃密、略結實和蛋糕孔很小，口感比較Q彈，蛋味很濃。
This foaming method of cake-making is called Genoise. Texture is dense, bouncy and firm with less holes in the cake but stronger egg aroma.

材料 Ingredients

餅底

雞蛋{ 2個 }
糖{ 30克 }
麵粉{ 30克 }
酥油{ 10克 }

餡料

馬斯卡邦起士{ 250克 }
蛋黃{ 2個 }
糖{ 30克 }
植物性鮮奶油{ 150克，打發 }
動物性鮮奶油{ 100克，打發 }
吉利丁粉{ 10克 }
滾水{ 70克 }
椰子汁{ 100克 }
蜜桃{ 100克，一半，切粒 }

Base

(2) eggs
(30g) sugar
(30g) flour
(10g) shortening

Filling

(250g) mascarpone cheese
(2) egg yolks
(30g) sugar
(150g) non-dairy cream, whipped
(100g) whipping cream, whipped
(10g) gelatin
(70g) boiling water
(100g) coconut juice
(100g) peach, halved and diced

難度●●●低

Coconut Peach Cheesecake

椰子蜜桃起士蛋糕

做法 Directions

餅底

1. 將雞蛋和糖打發；麵粉過篩，慢慢加入蛋糊中，攪拌均勻。
2. 加入酥油，攪拌均勻，倒進8吋餅模裏，放入已預熱的烤箱，用180℃烤10分鐘。
3. 取出蛋糕，放涼，切成2片，待用。

餡料

1. 將滾水與吉利丁粉拌至溶解，備用。
2. 將糖和蛋黃打至奶白色，加入奶油乳酪攪拌滑順，再加入椰子汁和蜜桃攪拌滑順。
3. 然後加入已溶化的吉利丁粉，攪拌均勻，最後拌入已打發的鮮奶油。

組合

將餅底放入餅模中，再加入餡料，放入冰箱冰2小時使蛋糕凝固後再取出，用椰絲及蜜桃裝飾。

Base

1. Beat egg and sugar until it forms soft peak. Sift flour and add to beaten egg. Mix well.
2. Add shortening. Combine. Pour in an 8-inch cake mould. Bake in a preheated oven at 180°C for 10 minutes.
3. Remove base cake from the oven. Cut in 2 slices. Set aside.

Filling

1. Blend gelatin with boiling water until it dissolves. Set aside.
2. Beat sugar and egg yolk until the colour turns to creamy white. Add cheese. Blend until smooth Add coconut juice and peach. Blend until smooth.
3. Then add gelatin solution. Combine. Finally fold in whipped cream.

Assembly

Put base cake in a cake mould. Pour in filling. Refrigerate for 2 hours until it is firm. Remove from the mould. Decorate with shredded coconut and peach.

Q：蛋糕用酥油會不會沉澱，使蛋糕無法膨脹？
Would shortening make cakes hard to expand?

A：其實做海綿蛋糕使用的油，任何種類都可以，只要是液體即可。但在蛋糕拌入油時必須力道要輕和徹底攪拌均勻，才不會讓油沉澱使蛋糕膨脹發不起。
As long as it is fluid, oil of any kind can be used. You have to blend batter with oil gently and thoroughly so that oil does not sink to the bottom and cakes can expand.

材料 Ingredients

餅底
- 壓碎的瑪莉餅{150克}
- 奶油{80克}

餡料
- 奶油乳酪{250克}
- 糖{30克}
- 蛋黃{2個}
- 吉利丁粉{10克}
- 滾水{70克}
- 植物性鮮奶油{125克}
- 動物性鮮奶油{125克}
- 牛奶{100克}
- 碎核桃{50克}
- 香草精{數滴}

裝飾
- 巧克力{50克}
- 動物性鮮奶油{50克}

Base
- (150g) marie biscuit crumbs
- (80g) butter

Filling
- (250g) cream cheese
- (30g) sugar
- (2) egg yolks
- (10g) gelatin
- (70g) boiling water
- (125g) non-dairy cream
- (125g) whipping cream
- (100g) milk
- (50g) walnuts, chopped
- A few drops vanilla extract

Decoration
- (50g) chocolate
- (50g) whipping cream

難度 ●●● 低

Vanilla Walnut Cheesecake

香草核桃起士蛋糕

做法 Directions

裝飾

將巧克力和動物性鮮奶油用熱水隔水加熱。

餅底

將壓碎的瑪莉餅和奶油混合，攪碎，放進8吋餅模裏，抹平。

餡料

1. 將奶油乳酪、糖和蛋黃打發；將植物性鮮奶油和動物性鮮奶油混合，打發。
2. 將吉利丁粉與滾水拌至融合，煮至溶化。
3. 將牛奶、碎核桃和香草精攪拌均勻。
4. 將奶油乳酪與已溶化的吉利丁粉混合，打至滑順，再加入牛奶核桃糊，打滑，最後拌入已打發的鮮奶油，變成餡料。

組合

在餅模裏，放入蛋糕片，倒入餡料，放入冰箱冰2小時使蛋糕凝固後再取出，用巧克力鮮奶油裝飾。

Decoration

Put chocolate and whipping cream in a container. Warm up in hot water until melted and combined.

Base

Combine maire biscuit crumbs and butter. Mix well. Put in and 8-10 inch cake mould. Press to level the surface.

Filling

1. Whip cream cheese, sugar and egg yolk. Combine non-dairy cream with whipping cream; beat.
2. Stir gelatin and boiling water until combined and dissolved.
3. Combine milk with chopped walnut and vanilla extract.
4. Combine cream cheese mixture with gelatin solution. Blend until smooth. Then add walnut milk; blend until smooth. Finally fold in whipped cream until it forms filling.

Assembly

Put cake slice in a cake mould. Pour in filling. Refrigerate for 2 hours until it is firm. Remove from the mould. Decorate with chocolate cream.

Q：坊間裏的巧克力有調溫與不調溫，應該用哪一種比較好？
Some chocolate is for tempering while some not. Which one should I use?

A：對於初學者，應該用不調溫的巧克力比較好，價錢實惠又容易處理，無論味道和效果都不錯。
Beginners should use the chocolate not for tempering as it is cheaper and easier to handle.The outcome will not be disappointing.

材料 Ingredients

手指餅蛋糕
- 蛋白{3個}
- 糖{50克}
- 蛋黃{3個}
- 麵粉{60克}

咖啡糖水
- 咖啡{1茶匙}
- 滾水{100克}
- 糖{20克}
- 蘭姆酒{少許}

馬斯卡邦起士餡
- 蛋黃{3個}
- 糖{80克}
- 馬斯卡邦起士{250克}
- 動物性鮮奶油{300克，打發}
- 吉利丁粉{10克}
- 滾水{70克}
- 咖啡酒{20克}

裝飾
- 可可粉{適量，灑蛋糕表面用}

....

Lady finger cake
- (3) egg whites
- (50g) sugar
- (3) egg yolks
- (60g) flour

Coffee syrup
- (1) tsp coffee
- (100g) boiling water
- (20g) sugar
- Rum

Mascarpone cheese filling
- (3) egg yolks
- (80g) sugar
- (250g) mascarpone cheese
- (300g) whipping cream, whipped
- (10g) gelatin
- (70g) boiling water
- (20) coffee liqueur

Decoration
- Cocoa powder, sprinkle over top

Tiramisu

提拉米蘇

做法 Directions

1. 手指餅蛋糕：將蛋白與糖打至濃稠，加入蛋黃攪拌均勻，加入麵粉攪拌滑順，放入擠花袋擠出長條狀，用180℃烤10-15分鐘。
2. 馬斯卡邦起士餡：將蛋黃和糖打至滑順，加入馬斯卡邦起士攪拌至柔滑融合，再倒入已調勻的吉利丁咖啡酒混合物攪拌均勻，慢慢拌入已打發的動物性鮮奶油。
3. 咖啡糖水：所有材料攪拌均勻。

組合

在已墊手指餅蛋糕的表面上，灑上咖啡糖水，然後倒入起士餡，放入冰箱冰2小時使蛋糕凝固後再取出，灑上可可粉即可。

1. For lady finger cake:Beat egg white and sugar until thickened. Stir in egg yolk. Add flour. Blend until smooth. Pour in an 8-inch cake mould. Bake at 180°C for 10-15 minutes.
2. For mascarpone cheese filling:Beat egg white and sugar until smooth. Add mascarpone cheese. Blend until smooth and combined. Then pour in coffee liqueur gelatin solution. Mix well. Slowly fold in whipped cream.
3. For coffee syrup:Combine all the ingredients.

Assembly

Brush the surface of lady finger cake with coffee syrup. Then pour in cheese filling. Refrigerate for 2 hours until it is firm. Remove from the mould. Sprinle with cocoa powder. Serve.

Q：傳說真正的提拉米蘇是用湯匙舀起享用，不是香港式的蛋糕，是真的嗎？
I heard that, originally, Italians eat tiramisu with a spoon, not with fingers like us. Is it true?

A：沒錯，真正的提拉米蘇是舀起來吃，酒味很濃，入口即化，不是整個蛋糕形式的表現。我覺得無論任何形式出現，只要選對自己的風格和味道，就可以食用。
Yes, they use a spoon to taste it. Strong liquor flavoured tiramisu should be melted in mouth. I don't mind either way how to eat tiramisu as long as you enjoy the style and taste.

材料 Ingredients

餅底

壓碎的瑪莉餅{150克}
奶油{80克}

餡料

奶油乳酪{250克}
糖{20克}
蛋黃{2個}
植物性鮮奶油{125克，打發}
動物性鮮奶油{125克，打發}
吉利丁粉{10克}
滾水{70克}
牛奶{120克}

裝飾

巧克力{50克}
動物性鮮奶油{50克}

Base

(150g) marie biscuit crumbs
(80g) butter

Filling

(250g) cream cheese
(20g) sugar
(2) egg yolks
(125g) non-dairy cream, whipped
(125g) whipping cream, whipped
(10g) gelatin
(70g) boiling water
(120g) milk

Decoration

(50g) chocolate
(50g) whipping cream

難度●●●低

Marble Cheesecake

雲石起士蛋糕

做法 Directions

餅底

將壓碎的瑪莉餅和奶油混合攪拌均勻，放進8吋餅模裏，抹平。

餡料

1. 將奶油乳酪攪拌滑順；糖和蛋黃打發，直至呈奶白色；用熱水將吉利丁粉溶解。
2. 煮滾牛奶，先加入糖蛋黃混合物攪拌均勻，再加入奶油乳酪攪拌滑順，最後加入已溶化的吉利丁粉攪拌均勻。
3. 將兩種已打發的鮮奶油攪拌均勻，慢慢拌入起士混合物攪拌均勻。

組合

將餡料放在餅底上，放入冰箱存放約2小時後取出。

裝飾

將50克動物性鮮奶油和巧克力用熱水隔水加熱，淋在起士蛋糕上。

Base

Base Combine marie biscuit crumbs with butter. Pit in an 8-inch cake mould. Press and level the surface.

Filling

1. Blend cream cheese until smooth; beat sugar and egg yolk until the colour turns to creamy white. Dissolve gelatin in hot water.
2. Bring milk to a boil. Add egg mixture. Combine. Then add cream cheese. Blend until smooth. Finally stir in gelatin solution.
3. Combine non-dairy cream and whipping cream, both whipped in advance. Slowly fold in cheese mixture until combined.

Assembly

Pour in filling over the base. Refrigerate for about 2 hours. Remove from the mould.

Decoration

Combine 50g whipping cream and chocolate in a container. Warm up in hot water until melted and combined. Drizzle over the cheesecake.

Q：雲石起士蛋糕是許多烘焙班必學的課程，這個概念有點像在咖啡裏拉花，可以自己發揮嗎？
Marble cheesecake is a must-have recipe in the curriculum of beginner's class. It's like making patterns in the foam topping of coffee. Can I be creative?

A：可以。在烘焙業裏已是經常被運用的雙色技巧，通常會把淺色糕糊做底色，另以深色作繪製圖紋，才能突顯圖案，在畫花紋時的要快和圖中已存在圖案樣式，加上糕糊比較濃稠，不易化開，才能顯現出效果。
Sure. It is a commonly used skill of culinary art. Chefs often use light batter as the base colour and touch it up with dark batter to create prominent contrast in patterns. It has to be quick and have the patterns in advance. Batter should also be thick to prevent colour from fading.

材料 Ingredients

餅底

巧克力蛋糕{1片}(參閱第121頁)
海綿蛋糕{1片}(參閱第121頁)

餡料

奶油乳酪{250克}
糖{20克}
蛋黃{2個}
牛奶{120克，煮熱}
植物性鮮奶油{150克，打發}
動物性鮮奶油{100克，打發}
吉利丁粉{8克}
滾水{50克}
蘭姆酒{10克}
白巧克力{50克}
動物性鮮奶油
{50克，白巧克力用}
黑巧克力{50克}
動物性鮮奶油
{50克，黑巧克力用}

裝飾

黑巧克力碎片{適量}
白巧克力碎片{適量}

Base

(1)chocolate cake slice
(Refer to p.121)

(1) slice base cake (Refer to p.121)

Filling

(250g) cream cheese

(20g) sugar

(2) egg yolks

(120g) milk, warm

(150g) non-dairy cream, whipped

(100g) whipping cream, whipped

(8g) gelatin

(50g) boiling water

(10g) rum

(50g) white chocolate

(50g) whipping cream, for white chocolate

(50g) dark chocolate

(50g) whipping cream, for dark chocolate

Decoration

Dark chocolate, chopped

White chocolate, chopped

難度●●●低

Chocolate Cheesecake

黑白巧克力起士蛋糕

做法 Directions

餡料

1. 糖和蛋黃打發，並加入熱牛奶攪拌均勻。
2. 奶油乳酪在室溫下放軟攪拌滑順，與蛋黃混合物攪拌均勻，將熱水與吉利丁粉加熱溶解，加入蘭姆酒攪拌均勻。
3. 將已混合的兩種鮮奶油打發，與蛋黃起士混合物攪拌均勻，分成兩份。
4. 白巧克力與黑巧克力各自與動物性鮮奶油混合，隔水加熱拌至溶解。
5. 黑巧克力漿與白巧克力漿，分別與一份起士混合物攪拌均勻。

組合

把一片巧克力蛋糕墊在餅模內，倒入黑巧克力起士混合物，再放上一片海綿蛋糕，倒入白巧克力起士混合物，放入冰箱冰1小時後，再從冰箱取出裝飾。

Filling

1. Beat sugar and egg yolk. Add warm milk and combine.
2. Leave cream cheese to be softened. Blend until smooth. Combine with egg yolk mixture. Dissolve gelatin in hot water. Stir in rum.
3. Combine mixed cream (non-dairy cream and whipping cream) with egg yolk cheese mixture. Divide in 2 portions.
4. Combine white chocolate and dark chocolate each with whipping cream in two containers respectively. Warm up in hot water until melted.
5. Combine dark chocolate cream and white chocolate cream each with 1 portion of cheese mixture respectively.

Assembly

Put a slice of chocolate cake in a cake mould. Pour in dark chocolate cheese mixture. Place a base cake slice on top. Pour in white chocolate cheese mixture. Refrigerate for 1 hour until it is firm. Remove from the mould. Decorate.

Q：巧克力的過程，為何要添加動物性鮮奶油？
Why should I add whipping cream when melting chocolate?

A：這是法國式做法，把巧克力與動物性鮮奶油一併溶解，稱為「Ganache」。它的特質是入口即化，形成柔軟、質如綿絨的濃郁巧克力醬，多用作糕果糕餅的餡心，也是巧克力慕斯的常用做法。
It's a French term, called Ganache, referring to melt chocolate and cream together. Chocolate Ganache is rich and silky chocolate spread, mainly used for cake filling or mousse.

材料 Ingredients

餅底

海綿蛋糕{2片}
(參閱第121頁)

餡料

奶油乳酪
{250克，攪拌滑順}
糖{40克}
蛋黃{2個}
牛奶{100克，煮熱}
植物性鮮奶油{150克}
動物性鮮奶油{100克}
吉利丁粉{10克}
滾水{70克}
蘭姆酒葡萄乾{100克}

裝飾

蛋黃{1個}
櫻桃{3粒}
巧克力片{3-5片}
蘭姆酒葡萄乾{適量}
鮮奶油，打發{100克}

Base

(2) slices base cake
(Refer to p.121)

Filling

(250g) cream cheese, blend to smooth
(40g) sugar
(2) egg yolks
(100g) milk, boil to warm
(150g) non-dairy cream
(100g) whipping cream
(10g) gelatin
(70g) boiling water
(100g) rum raisins

Decoration

(1) egg yolks
(3) cherries
(3-5) chocolate shavings
Rum raisins
(100g) cream, whipped

難度●●●低

Germany Rum Cheesecake

德國蘭姆酒起士蛋糕

做法 Directions

餡料

1. 將動物性鮮奶油和植物性鮮奶油混合，打發。
2. 將糖和蛋黃打至奶白色，再倒入熱牛奶攪拌均勻，與已打滑順的奶油乳酪混合。
3. 滾水和吉利丁粉混合，在熱水隔水加熱拌至溶解，與起士混合物攪拌均勻，再加入蘭姆酒葡萄乾攪拌均勻。慢慢拌入已打發奶油。

組合

1. 將一片蛋糕放入餅模內，倒進半份餡料，再放一片蛋糕，倒入剩餘的餡料，放入冰箱凝固2小時。
2. 取出蛋糕，抹上已打發的鮮奶油，在蛋糕表面塗蛋黃液，用噴火槍噴燒已塗蛋黃液的蛋糕表面，再放上其他裝飾。

Filling

1. Combine non-dairy cream and whipping cream; beat.
2. Beat sugar and egg white until the colour turns to creamy white. Pour in warm milk. Blend. Combine with whipped cream cheese.
3. Combine boiling water and gelatin in a container. Warm up in hot water until dissolved. Combine with cheese mixture. Stir in rum raisins. Slightly fold in whipped cream.

Assembly

1. Put a cake slice in a cake mould. Pour in 1/2 portion of the Filling. Place another cake slice on top. Pour the remainder of the Filling, Refigerate for 2 hours intil it is firm.
2. Remove the cake from the mould. Coat with whipped cream. Brush the surface with whisked egg yolk. Burn with a torch. Decorate with other ingredients.

Q：蘭姆酒葡萄乾可以自己做嗎？
Can I preserve rum raisins myself?

A：蘭姆酒葡萄乾的風味很好，與奶油或鮮奶油類材料，十分匹配，做法很簡單，用葡萄乾50克配蘭姆酒125克浸泡一星期，葡萄乾會完全吸入蘭姆酒香味和水分，變脹大和柔軟，如時間不許可，至少要浸泡一夜，才會有好效果。
Rum raisins go well with butter or cream. It is way easy to preserve: soak 50g raisins in 125g rum for a week. Raisins expand and become tender by absorbing the aroma and water of rum. If time is limited, soak them overnight at least.

Q：我可以在芒果餡料加入50克芒果粒，增加口感嗎？
Can I add 50g mango cubes in the filling to enhance the consistency?

A：當然可以，全用芒果果泥做餡料，口感順滑細緻，味道濃郁，要是加入芒果肉小粒，使進食層次有對比，但切記不要加太多而使夾餡碎裂。
Sure. Mango puree enriches the flavour while cubes give some contrast. But remember not to add too much or it will break the jelly layer.

難度●●●中

Mango Mousse Cheesecake

芒果慕斯起士蛋糕

材料 Ingredients

派底

海綿蛋糕{1片}
(參閱第121頁)

餡料

奶油乳酪{125克}
糖{40克}
動物性鮮奶油
{300克，打發}
芒果果泥{100克}
牛奶{50克}
吉利丁粉{12克}
滾水{70克}

裝飾

芒果切片放在蛋糕表面

Base

(1) slice base cake
(Refer to p.121)

Filling

(125g) cream cheese
(40g) sugar
(300g) whipping cream, whipped
(100g) mango puree
(50g) milk
(12g) gelatin powder
(70g) boiling water

Decoration

mango slices

做法 Directions

餡料

1. 將奶油乳酪在室溫放軟備用，再與糖混合攪拌滑順。
2. 滾水與吉利丁粉攪拌均勻，在熱水隔水加熱拌至溶解。
3. 將已溶化的吉利丁粉、芒果果泥和牛奶同置碗中攪拌均勻，加入奶油乳酪混合物中攪拌均勻，慢慢加入已打發的鮮奶油攪拌滑順。

組合

1. 在8吋餅模內，放入海綿蛋糕片，倒進餡料，放入冰箱冰1-2小時後，從冰箱取出。
2. 取出蛋糕，在蛋糕表面倒入芒果果凍，放回冰箱冷凍至凝固，再從冰箱取出裝飾。

Filling

1. Leave cream cheese to soften at room temperature. Combine with sugar and whisk to smooth.
2. Combine boiling water with gelatin powder in hot water. Blend until dissolved.
3. In a bowl, combine gelatin solution with mango puree and milk. Gradually fold in whipped cream cheese and whipped cream according. Blend until smooth.

Assembly

1. Put a base cake slice in an 8 inch cake mould. Pour in Filling. Refrigerate for 1-2 hours until it is firm.
2. Take out the cake. Pour mango jelly over the top. Refrigerate again until it is firm. Take out and decorate.

Q：把麵糰按壓於派模，有沒有技巧要求？

What is the skill of pressing dough into the mould?

A：基本上是沒有固定技巧，只要把它均勻地按入派模內，或是把麵糰桿薄，覆蓋於派模，再用刮刀削去多餘麵皮。建議要把麵糰搓揉後置冰箱冷凍15-20分鐘等硬實一點，較容易展開麵糰。

Evenly press it into the mould. Roll it into a thin layer and cover the mould. Use a knife to cut off the overhang. You are advised to chill dough for 15-20 minutes until it is slightly firm, to roll out easily.

難度●●●中

Tomato and Asparagus Cheese Pie

香茄鮮蘆筍起士派

材料 Ingredients

派底

奶油{220克}
雞蛋{1個}
麵粉{360克}
糖{80克}

餡料

番茄{2個}
蘆筍{50克}
Pizza起士{50克}
雞蛋{2個}
牛奶{200克}
鹽{1克}
黑胡椒粉{少許}

Pie crust

(220g) butter
(1) egg
(360g) flour
(80g) sugar

Filling

(2) tomatoes
(50g) asparagus
(50g) mozzarella cheese pizza
(2) eggs
(200g) milk
(1g) salt
Black pepper

做法 Directions

派底

1. 麵粉過篩，並在麵粉中央用麵粉築起粉牆。
2. 放進剩餘材料，搓揉成軟滑麵糰，置於冰箱冷凍20-30分鐘。
3. 取出派皮麵糰，桿成約3毫米厚，放進8吋餅模裏，輕輕壓實。

餡料

1. 番茄洗淨，切丁。蘆筍洗淨、切丁。
2. 牛奶和雞蛋輕輕打至均勻，並加入黑胡椒粉和鹽調味。

組合

1. 將番茄丁和蘆筍丁放進已鋪上派皮糰的派模上，再注入蛋奶糊，派面撒上pizza起士。
2. 烤箱預熱，放入起士派，用200℃烤15-20分鐘。

Pie crust

1. Sieve flour. Make a hole in the centre of the pile of flour.
2. Add the remainder of ingredients. Knead and form to smooth dough. Refrigerate for 20-30 minutes.
3. Take out the dough. Roll out into 3 mm thick. Put in an 8 inch cake mould. Gently press it with fingertips.

Filling

1. Rinse tomatoes and cut in cubes. Rinse asparagus and cut in cubes.
2. Gently beat milk and egg. Add black pepper and salt to season.

Assembly

1. Place tomato and asparagus cubes on the pie crust. Pour in beaten egg milk. Sprinkle with pizza over the top.
2. Bake in a preheated oven at 200°C for 15-20 minutes.

材料 Ingredients

派底

奶油{220克}
雞蛋{1個}
麵粉{360克}
糖{80克}

餡料

糖{40克}
雞蛋{1個}
奶油乳酪
{100克，放軟攪拌滑順}
檸檬皮{1/2個}
動物性鮮奶油{50克}
麵粉{1/2茶匙}

紅酒煮洋梨

紅酒{70克}
糖{15克}
肉桂枝{2克}
檸檬汁{少許}
去皮洋梨{2個}

Pie crust

(220g) butter, (1) egg
(360g) flour, (80g) sugar

Filling

(40g) sugar, (1) egg
(100g) cream cheese, softened and beaten to smooth
(1/2) lemon zest
(50g) whipping cream
(1/2) tsp flour

Pear paoched in red wine

(70g) red wine, (15g) sugar
(2g) cinnamon stick
Lemon juice
(2) pears, peeled

Red Wine Pear Cheese Pie

紅酒洋梨起士派

做法 Directions

派底

1. 將麵粉過篩，再用麵粉築起粉牆，放進剩餘材料，搓揉成軟滑麵糰，置於冰箱冰20-30分鐘。
2. 取出派皮麵糰，桿成約3毫米厚，放進8吋餅模裏，輕輕壓實。

餡料

將雞蛋與糖打發，與奶油乳酪攪拌均勻，再與其餘餡料攪拌均勻，混合成起士麵糊。

紅酒煮洋梨

將洋梨切片，再和其他材料同置鍋中一起煮滾，靜放2小時，備用。

組合

取出派底，放上洋梨片，再倒進起士麵糊，放入已預熱的烤箱內，用180℃烤約25-30分鐘。

Pie crust

1. Sieve flour. Make a hole in the centre of the pile of flour. Add the remainder of ingredients. Knead and form to smooth dough. Refrigerate for 20-30 minutes.
2. Take out the dough. Roll out into 3 mm thick. Put in an 8 inch cake mould. Gently press it with fingertips. with fingertips.

Filling

Beat egg yolk and sugar. Combine with cream cheese and blend. Combine the remainder of ingredients and blend to cheese batter.

Pear paoched in red wine

Slice pear. In a saucepan, combine with the remainder of ingredients and bring to a boil. Set aside for 2 hours.

Assembly

Take out the pie crust. Place pear slices on top. Pour in cheese batter. Bake in a preheated oven at 180°C for 25-30 minutes.

Q：為什麼有時需要在派底刺孔，有時又不需要，究竟那樣才是對的呢？
How come it is not necessary to poke the pie crust in some recipes?

A：派底是否需要刺孔，決定於盛載餡料的質感和有沒有必要同時入烤箱烘烤，如果是流質餡料或是連同餡料一起烘烤，就無需刺孔，否則會流出來黏在模具上；要是烘烤後才添加餡料，就要先刺孔再入烤箱烘焙。
It depends on the consistency of filling and if the curst is baked together. If the filling is fluidly and bake together in the curst, you don't need to poke it; otherwise, it spills over and sticks to the mould. If you add filling after the curst is baked, you have to poke it before you put it in an oven.

材料 Ingredients

派底

奶油{ 220克 }
雞蛋{ 1個 }
麵粉{ 360克 }
糖{ 80克 }

餡料

奶油乳酪{ 500克，放軟 }
蛋黃{ 4個 }
糖{ 80克 }
蛋白{ 4個 }
白蘭地酒{ 120克 }
櫻桃{ 50克 }

Pie crust

(220g) butter
(1) egg
(360g) flour
(80g) sugar

Filling

(500g) cream cheese, softened
(4) egg yolks
(80g) sugar
(4) egg whites
(120g) brandy
(50g) cherries

難度●●●中

Brandy Cheery Cheese Pie

洋酒櫻桃起士派

做法 Directions

派底

1. 將麵粉過篩，再用麵粉築起粉牆，放進剩餘材料，搓揉成軟滑麵糰，置於冰箱冷凍20-30分鐘。
2. 取出派皮麵糰，桿成約3毫米厚，放進8吋餅模裏，輕輕壓實。

餡料

1. 將奶油乳酪和蛋黃攪拌滑順。
2. 打發蛋白和糖，慢慢拌入奶油乳酪蛋糊裏，混合。
3. 將白蘭地酒和櫻桃混合，攪拌均勻，放入奶油乳酪麵糊裏，打發成餡料。

組合

1. 取出派底，放入已預熱的烤箱，以190℃烘烤派底，取出。
2. 放進適量餡料，回放烤箱中用170℃烤45分鐘即可。

小提醒

可用蘭姆酒或櫻桃酒取代白蘭地酒。

Pie crust

1. Sieve flour. Make a hole in the centre of the pile of flour. Add the remainder of ingredients. Knead and form to smooth dough. Refrigerate for 20-30 minutes.
2. Take out the dough. Roll out into 3 mm thick. Put in an 8 inch cake mould. Gently press it with fingertips.

Filling

1. Combine cream cheese and egg yolk. Blend until smooth.
2. Beat egg white and sugar. Gradually fold in cream cheese batter. Combine.
3. Combine brandy with cherries. Add to cream cheese. Blend and form to Filling.

Assembly

1. Take out pie crust. Bake in a preheated oven at 190°C until cooked. Take out.
2. Add Filling. Bake again in the oven at 170°C for 45 minutes. Done.

Tip

Rum or cherry liquor may replace brandy.

Q：派底預先烤熟，與連同餡料生烤的分別為何？
What is the difference between baking crust in advance and baking crust and filling together?

A：派底預先烤熟，能確保派底酥脆和完全熟透；連同餡料生烤，要是派底太厚和底火不足，往往會出現未完全熟的狀況，做不了香脆效果。不過最主要都是看製品的效果和需要而決定。
Bake crust in advance is to keep its crispness and make sure it is cooked thoroughly. If it is baked together with filling, curst may not be well done due to its excessive thickness or insufficient low heat. It may spoil the crispy texture. But in most cases, it depends on what texture you wanted to achieve.

Q：坊間有冷凍、新鮮和罐裝藍莓醬，究竟用哪一種比較適合？
Among frozen, fresh and canned blueberry jam, which one should I pick?

A：沒有一定的準則。冷凍藍莓整顆入冰，所以解凍後含水分較多，味道淡一點也比較柔軟，一年四季皆有。新鮮水果就有季節性，不過近年就因來源地多了，供應穩定。至於藍莓醬屬糖煮加工製品，方便省時，味道帶甜，按自己喜好和習慣選用。
It depends. They freeze the whole blueberries so it is more juicy and plain in taste. It is convenient to get it during the year. Fresh blueberries are seasonal but, with more sources, supplies are stable. Blueberry jam saves time and brings sweetness to the cake.

Blueberry Cheesecake
藍莓起士蛋糕

材料 Ingredients

餅底
- 海綿蛋糕{2片}（參閱第121頁）
- 藍莓醬{適量，裝飾用}

餡料
- 奶油乳酪{250克，放軟}
- 糖{20克，奶油乳酪用}
- 蛋黃{2個}
- 糖{20克，蛋黃用}
- 吉利丁粉{12克}
- 滾水{70克}
- 植物性鮮奶油{150克，已打發}
- 動物性鮮奶油{100克，已打發}
- 藍莓醬{20克}

Base
- (2) slices base cake (Refer to p.121)
- blueberry jam, for Decoration

Filling
- (250g) cream cheese, softened
- (20g) sugar, for cream cheese
- (2) egg yolks
- (20g) sugar, for egg yolk
- (12g) gelatin powder
- (70g) boiling water
- (150g) non-dairy cream, whipped
- (100g) whipping cream, whipped
- (20g) blueberry jam

做法 Directions

餡料
1. 將奶油乳酪和20克糖攪拌滑順；將蛋黃和20克糖攪拌滑順。
2. 用滾水將吉利丁粉煮溶。
3. 將奶油乳酪糊和蛋黃糊同放一起攪拌滑順。
4. 然後倒入已溶化的吉利丁粉攪拌滑順，慢慢拌入已打發的鮮奶油，再加入藍莓醬輕輕混合。
5. 備一蛋糕模，放入一片海綿蛋糕，倒入半份餡料，再放一片海綿蛋糕，然後倒入剩餘餡料，置於冰箱冰1-2小時後取出，擠上藍莓醬裝飾。

Filling
1. Blend cream cheese and 20g sugar to smooth. Blend egg yolk and 20g sugar to smooth.
2. Dissolve gelatin powder in boiling water.
3. Blend sweetened cream cheese (1) and sweetened egg yolk (1).
4. Pour in gelatin solution. Blend until smooth. Gradually fold in whipped cream. Add blueberries jam. Gently combine.
5. Put a base cake slice into a cake mould. Pour in half of Filling. Place another base cake slice on top. Pour in the remainder of Filling. Refrigerate for 1-2 hours. Take out. Pipe blueberry jam for decoration.

材料 Ingredients

餅底

海綿蛋糕{1片}
（參閱第121頁）

餡料

奶油乳酪{125克，放軟}
糖{20克，奶油乳酪用}
蛋黃{2個}
糖{20克，蛋黃用}
吉利丁粉{12克}
滾水{70克}
植物性鮮奶油{150克}
動物性鮮奶油{100克}
芒果果泥{80克}
葡萄柚果肉{30克}
熟西谷米{60克}

果凍

果凍粉{10克}
滾水{200克}
糖{20克}
芒果果泥
{20克，最後才加}

Base

(2) slices base cake
(Refer to p.121)

Filling

(125g) cream cheese, softened
(20g) sugar, for cream cheese
(2) egg yolks
(20g) sugar, for egg yolk
(12g) gelatin powder
(70g) boiling water
(150g) non-dairy cream
(100g) whipping cream
(80g) mango puree
(30g) grapefruit pulp
(60g) sago, cooked

Jelly

(10g) jelly powder
(200g) boiling water
(20g) sugar
(20g) mango puree, add at last

Pomelo Mango Cheesecake

楊枝甘露起士蛋糕

做法 Directions

1. 將奶油乳酪和20克糖攪拌溶化。把蛋黃和20克糖打至滑順。
2. 用滾水將吉利丁粉煮至溶化。
3. 再將動物性鮮奶油和植物性鮮奶油混合，打發。
4. 將鮮奶油糊和蛋糊混合，再加入已溶化的吉利丁粉攪拌均勻，慢慢拌入已打發的鮮奶油。
5. 備一蛋糕模，放入一片海綿蛋糕，倒入半份餡料，再放一片海綿蛋糕，然後倒入剩餘餡料，置於冰箱冰1-2小時。
6. 果凍材料(除芒果果泥外)攪拌均勻，取出起士蛋糕，倒進果凍材料，放回冰箱冰1-2小時即可。

1. Blend cream cheese and 20g sugar to smooth. Blend egg yolk and 20g sugar to smooth.
2. Dissolve gelatin powder in boiling water.
3. Combine non-dairy cream and shipping cream. Beat.
4. Blend sweetened cream cheese (1) and sweetened egg yolk (1). Pour in gelatin solution. Combine. Gradually fold in whipped cream.
5. Put a base cake slice into a cake mould. Pour in half of Filling. Place another base cake slice on top. Pour in the remainder of Filling. Refrigerate for 1-2 hours.
6. Combine ingredients of jelly (except mango puree). Take out cheesecake. Pour in jelly ingredients. Refrigerate again for 1-2 hours. Done.

Q：芒果應該選生、半熟還是全熟？
How ripe is mango for the recipe?

A：未成熟的芒果，果肉硬實、沒香味和很酸；半熟的芒果，質感柔軟而堅挺，但味道不夠香甜；全熟的芒果，香味足而甜美，但是果肉太柔軟，如果純打作果泥沒問題，要是切粒或切片就不易成型。建議打果泥取全熟，表皮有黑點或皺皮都可以，最重要有芒果香味。
Unripe mango is firm, sour without aroma. Half-ripe mango turn to tender with substantial firmness, but taste is not sweet enough. When it is sweet and aromatic, the texture is too tender. To make puree, you are suggested to blend ripe mangoes, with dark sports on the skin or wrinkled skin.

Q：低脂起士是甚麼？

What is low fat cream cheese?

A：它是以多種碳水化合物或蛋白質取代脂肪的起士製品，味道清淡，脂肪含量比較低，質感也比較柔軟，口感輕盈，對於愛吃起士又怕肥胖的人士，是不錯的選擇。

It is a process cheese product, which various carbohydrates and proteins are used to replace fat. Taste is light and texture is tender with lower content of fat. It is a good choice for those who care for weight managemen.

Low Fat Mixed Fruit Yogurt Cheesecake
鮮果優格低脂起士蛋糕

材料 Ingredients

餅底
海綿蛋糕{2片}
（參閱第121頁）

餡料
低脂起士{125克，放軟}
鮮果優格{200克}
吉利丁粉{10克}
滾水{70克}
植物性鮮奶油{150克，打發}
動物性鮮奶油{100克，打發}

Base
(2) slices base cake
(Refer to p.121)

Filling
(125g) low fat cheese, softened
(200g) fresh fruit yogurt
(10g) gelatin powder
(70g) boiling water
(150g) non-dairy cream, whipped
(100g) whipping cream, whipped

做法 Directions

餡料
1. 將低脂起士和鮮果優格混合。
2. 用滾水將吉利丁粉溶解，要是未能完全溶解，可隔水加熱繼續攪拌至清澈。
3. 將起士優格醬和已溶化的吉利丁粉混合，攪拌滑順，慢慢拌入鮮奶油。
4. 備一蛋糕模，放入一片海綿蛋糕，倒入半份餡料，再放一片海綿蛋糕，然後倒入剩餘餡料，放入冰箱冰1-2小時。
5. 取出，用植物性鮮奶油和水果裝飾。

Filling
1. Combine low fat cheese and fresh fruit yogurt.
2. Dissolve gelatin powder in boiling water. If it does not dissolve completely, warm it up in hot water, stirring until it is clear.
3. Combine yogurt cheese mixture and gelatin solution, blend until smooth. Gradually in fold in cream.
4. Put a base cake slice into a cake mould. Pour in half of Filling. Place another base cake slice on top. Pour in the remainder of Filling. Refrigerate for 1-2 hours.
5. Take out. Garnish with non-dairy cream and fruits.

Q：這個蛋糕可以用新鮮栗子做餡嗎？
Can I use fresh chestnut instead?

A：我初入行時，現成的烘焙材料不多，價錢比較貴，為了節省成本，我們會買新鮮栗子烘焙45分鐘，去殼，磨成泥，再加清水、牛奶和糖煮至柔順，最後加入奶油和蘭姆酒攪拌均勻，過篩放涼即可。
You may bake chestnuts for 45 minutes. Remove shells and mash. Combine with water, milk and sugar. Cook until tender. Combine with butter and rum. Sieveand form to puree.

Chestnut Mousse Cheesecake

栗子慕斯起士蛋糕

材料 Ingredients

餅底

海綿蛋糕{2片}
(參閱第121頁)
栗子泥{適量，裝飾用}
水果{適量，裝飾用}

餡料

奶油乳酪{100克}
栗子泥{100克}
蘭姆酒{10克}
吉利丁粉{12克}
滾水{70克}
牛奶{100克}
植物性鮮奶油
{150克，打發}
動物性鮮奶油
{100克，打發}

Base

(2) slices base cake
(Refer to p.121)
Chestnut puree, for Decoration
Fruits, for Decoration

Filling

(100g) cream cheese
(100g) chestnut puree
(10g) rum
(12g) gelatin powder
(70g) boiling water
(100g) milk
(150g) non-dairy cream, whipped
(100g) whipping cream, whipped

做法 Directions

餡料

1. 將奶油乳酪、栗子泥和蘭姆酒混合，攪拌滑順。
2. 將滾水和牛奶煮熱，放進吉利丁粉，煮溶。
3. 將栗子糊和已溶化的吉利丁粉混合，攪拌滑順，拌入已打發的鮮奶油。
4. 備一蛋糕模，放入一片海綿蛋糕，倒入半份餡料，再放一片海綿蛋糕，然後倒入剩餘餡料，放入冰箱冰1-2小時，
5. 取出起士蛋糕，然後用栗子和水果裝飾。

Filling

1. Combine cream cheese, chestnut pure and rum. Blend until smooth.
2. Warm boiling water and milk. Add gelatin powder. Cook until dissolved.
3. Combine chestnut mixture and gelatin. Blend until smooth. Fold in whipped cream.
4. Put a base cake slice into a cake mould. Pour in half of Filling. Place another base cake slice on top. Pour in the remainder of Filling. Refrigerate for 1-2 hours.
5. Take out the cheesecake. Garnish with chestnut and fruits.

Q：用哪一種芋頭比較好？
What kind of taro is better for this recipe?

A：中國品種和泰國品種各有所長，只要它夠粉糯，就是澱粉含量足，有芋香即可。選取時以較輕，尾端的莖位不大，切開後很快出現粉末狀和紫絲，就是好的芋頭。
Pick lightweight and starchy taros with thin ends. It should be powdery inside with purple threads when you cut it.

難度●●●中

Taro Coconut Cheesecake

香芋椰子起士蛋糕

材料 Ingredients

餅底

海綿蛋糕{2片}
(參閱第121頁)
椰泥{適量，裝飾用}
巧克力{適量，裝飾用}

餡料

奶油乳酪{125克}
椰子汁{100克}
香芋泥{100克}
吉利丁粉{12克}
滾水{70克}
植物性鮮奶油{150克，打發}
動物性鮮奶油{100克，打發}

Base

(2) slices base cake
(Refer to p.121)
Shredded coconut, for Decoration
Chocolate, for Decoration

Filling

(125g) cream cheese
(100g) coconut juice
(100g) taro puree
(12g) gelatin powder
(70g) boiling water
(150g) non-dairy cream, whipped
(100g) whipping cream, whipped

做法 Directions

1. 將奶油乳酪和椰子汁混合；用滾水將吉利丁粉煮溶；將動物性鮮奶油和植物性鮮奶油混合，攪拌均勻。
2. 先將椰子汁和香芋泥混合，攪拌均勻；再加入鮮奶油，攪拌滑順。
3. 備一蛋糕模，放入一片海綿蛋糕，倒入半份餡料，再放一片海綿蛋糕，然後倒入剩餘餡料，放入冰箱冰1-2小時。
4. 取出起士蛋糕，然後用椰泥和巧克力裝飾。

1. Combine cream cheese with coconut juice. Dissolve gelatin powder in boiling water. Combine non-dairy cream and whipping cream and blend.
2. Combine coconut juice with taro puree. Blend. Add cream and blend until smooth.
3. Put a base cake slice into a cake mould. Pour in half of Filling. Place another base cake slice on top. Pour in the remainder of Filling. Refrigerate for 1-2 hours.
4. Take out the cheesecake. Garnish with shredded coconut and chocolate.

材料 Ingredients

餅底

海綿蛋糕{2片}
（參閱第121頁）

餡料

奶油乳酪{250克，放軟}
糖{20克}
酸奶油{50克}
吉利丁粉{12克}
滾水{70克}
植物性鮮奶油{150克，打發}
動物性鮮奶油{100克，打發}
薑汁{20克}
自製糖薑{50克}

自製糖薑

糖{100克}
清水{90克}
薑片{50克}

Base

(2) slices base cake
(Refer to p.121)

Filling

(250g) cream cheese, softened
(20g) sugar
(50g) sour cream
(12g) gelatin powder
(70g) boiling water
(150g) non-dairy cream, whipped
(100g) whipping cream, whipped
(20g) ginger juice
(50g) homemade crystallized ginger

Homemade crystallized ginger

(100g) sugar
(90g) water
(50g) ginger slices

Ginger Cheesecake

薑味起士蛋糕

做法 Directions

自製糖薑

把所有材料煮至濃稠收汁。

餡料

1. 將奶油乳酪、糖和酸奶油放在一起混合，攪拌滑順。
2. 用滾水將吉利丁粉煮至溶化。
3. 將奶油乳酪混合物和已溶化的吉利丁粉混合，攪拌滑順，加入已混合的鮮奶油，攪拌均勻，再放入自製糖薑和薑汁攪拌均勻，做成餡料。

組合

1. 在蛋糕模裏放入一片海綿蛋糕，再倒入半份餡料，再蓋一片海綿蛋糕，倒入剩餘餡料，放進冰箱冰1-2小時。
2. 取出蛋糕，抹一層薄鮮奶油，沾上椰泥及巧克力裝飾。

Homemade crystallized ginger

Cook all the ingredients of homemade crystallized ginger until juice has evaporated.

Filling

1. Combine cream cheese with sugar and sour cream. Blend until smooth.
2. Dissolve gelatin powder in boiling water.
3. Combine cream cheese mixture with gelatin solution. Blend until smooth. Add combined cream. Combine. Add crystallized ginger and ginger juice. Combine and form to Filling.

Assembly

1. Put a base cake slice into a cake mould. Pour in half of Filling. Place another base cake slice on top. Pour in the remainder of Filling. Refrigerate for 1-2 hours.
2. Take out the cake. Spread a thin layer of cream. Garnish with shredded coconut and chocolate.

Q：想做糖薑片裝飾，應該如何處理？
How should I process crystallized ginger for use of decoration?

A：可以把自製糖薑放入已預熱的烤箱，用低溫火力160℃烘焙4-6小時至完全乾透，再放入密封盒內貯存。如允許，可放入乾燥劑保持乾爽和堅挺。
Bake crystallized ginger in a preheat oven at 160°C for 4-6 hours until it is dry completely. Store it up in a sealed box.

材料 Ingredients

餅底

海綿蛋糕{2片}
（參閱第121頁）

餡料

奶油乳酪{200克}
糖{20克}
豆腐{20克}
豆漿{100克}
吉利丁粉{12克}
滾水{70克}
鮮奶油{250克，打發}

裝飾

薄荷葉{1小棵}
豆腐起士球{數顆}
覆盆子{數顆}

Base

(2) slices base cake
(Refer to p.121)

Filling

(200g) cream cheese
(20g) sugar
(20g) tofu
(100g) soy milk
(12g) gelatin powder
(70g) boiling water
(250g) whipping cream, whipped

Decoration

(1) mint leaf
Tofu cheese balls
Raspberries

難度●●●中

Tofu Cheesecake

山水豆腐起士蛋糕

做法 Directions

餡料

1. 將奶油乳酪放軟，再與糖混合攪拌滑順。
2. 滾水與吉利丁粉拌至溶解。
3. 將奶油乳酪豆腐糊和吉利丁粉攪拌均勻，慢慢加入已打發的鮮奶油攪拌滑順。

組合

1. 將一片海綿蛋糕放進糕模裏，加入半份餡料，再放另一片海綿蛋糕，放入冰箱冰約1-2小時凝固後取出。
2. 在蛋糕表面上隨意放豆腐起士球、薄荷葉和覆盆子裝飾。

Filling

1. Combine cream cheese with sugar, tofu and soy milk.
2. Dissolve gelatin powder in boiling water.
3. Combine tofu cream cheese mixture with gelatin solution. Gradually foldi n whipped cream. Blend until smooth.

Assembly

1. Put a base cake slice into a cake mould. Pour in half of Filling. Place another base cake slice on top. Pour in the remainder of Filling. Refrigerate for 1-2 hours.
2. Garnish the cake with cheese tofu balls, mint leaf and raspberries.

Q：是不是用任何豆腐都可以？
Can I use tofu of any kind?

A：基本上任何豆腐都可以，在價錢有考量的話，可用市場賣的板豆腐，但要煨煮片刻，確保衛生；日本絹豆腐都是不錯選擇，有豆香兼水分少，但價錢略貴。
Yes, you may buy tofu in wet market but make sure to boil it briefly in water for hygiene reason. It is more expensive if you use Japanese silk tofu, which contains less water.

材料 Ingredients

餅底
消化餅底{1份}
（請參閱第122頁）
阿華田{10克}，灑蛋糕表面

餡料
馬斯卡邦起士{150克}
阿華田{40克}
滾水{20克，調阿華田用}
吉利丁粉{12克}
滾水{70克，調吉利丁粉用}
植物性鮮奶油
{100克，已打發}
動物性鮮奶油
{200克，已打發}

飾面蛋白霜
蛋白{60克}
糖{10克}
吉利丁粉{5克}
滾水{30克}

Base
(1) digestive biscuit base
(Refer to p.122)

(10g) Ovaltine,
sprinkle over the surface

Filling
(150g) mascarpone cheese

(40g) Ovaltine

(20g) boiling water,
add to Ovaltine

(12g) gelatin powder

(70g) boiling water,
add to gelatin powder

(100g) non-dairy cream, whipped

(200g) whipping cream, whipped

Italian meringue for Decoration
(60g) egg white

(10g) sugar

(5g) gelatin powder

(30g) boiling water

難度●●●中

Ovaltine Cheesecake

阿華田起士蛋糕

做法 Directions

餡料

1. 將阿華田和滾水混合攪拌融化，再與馬斯卡邦起士攪拌滑順。
2. 把滾水和吉利丁粉混合，煮溶。
3. 然後把馬斯卡邦起士糊和已溶化的吉利丁粉混合，拌入鮮奶油，攪拌均勻。

義大利蛋白霜

1. 用滾水將吉利丁粉煮溶。
2. 打發蛋白，加入糖打至滑順及硬性發泡，再倒入已溶化的吉利丁粉拌勻。

組合

1. 在8吋餅模裏放進消化餅底，輕輕壓實，鋪上餡料，放進冰箱冰2小時，使蛋糕凝固。
2. 取出起士蛋糕，抹上蛋白霜裝飾，再灑上阿華田粉裝飾即可。

Filling

1. Dissolve Ovaltine in boiling water. Combine with mascarpone cheese. Blend until smooth.
2. Dissolve gelatin powder in boiling water.
3. Combine mascarpone cheese mixture with gelatin solution. Fold in cream. Combine.

Italian meringue

1. Dissolve gelatin powder in boiling water.
2. Beat egg white. Add sugar. Beat until it is smooth and stiff. Add gelatin solution. Combine and form to meringue.

Assembly

1. Put digestive biscuit base in an 8 inch cake mould. Press it gently with fingertips. Pour in filling. Refrigerate for 2 hours until it is firm.
2. Take out the cheesecake. Spread over top meringue. Sprinkle with Ovaltine for Decoration. Done.

Q：蛋白霜(馬玲)有多少基本做法？
How many methods are used to make meringue?

A：蛋白霜大致上有三種做法：法式蛋白霜、瑞士蛋白霜和義式蛋白霜。它們基本上以蛋白和糖為主要原料，法式蛋白霜是簡單做法，只把蛋白打發，再分數次加入糖，打至濃稠；瑞士蛋白霜就把砂糖和蛋白放在容器，再置於瓦斯爐上以低溫將糖煮溶，取出繼續打至硬性發泡；義式蛋白霜就是把糖煮成糖糊，沖入已打發的蛋白，攪打至融和。
There are three:French, Swiss and Italian-styles. Basic ingredients are egg white and sugar. French style is the easiest one: beat egg white, gradually adding sugar at a few times, and beat until thickened. As for Swiss style, put sugar and egg white in a bowl. Warm it up over low heat to dissolve sugar. Take out and beat until stiff. To make Italian style meringue, heat sugar and it forms to syrup. Add beaten egg white in one gush. Beat until combined.

材料 Ingredients

餅底

巧克力蛋糕{1片}
(參閱第121頁)

海綿蛋糕{1片}
(參閱第121頁)

黑、白巧克力碎片
{適量，裝飾表面用}

餡料

馬斯卡邦起士{250克}

蛋黃{3個}

吉利丁粉{10克}

滾水{70克}

植物性鮮奶油
{150克，已打發}

動物性鮮奶油
{100克，已打發}

黑巧克力{50克，熱溶}

動物性鮮奶油
{50克，黑巧克力用}

Base

(1) slice chocolate cake
(Refer to p.121)

(1) slice base cake
(Refer to p.121)

Dark and white chocolate, grated for Decoration

Filling

(250g) mascarpone cheese

(3) egg yolks

(10g) gelatin powder

(70g) boiling water

(150g) non-dairy cream, whipped

(100g) whipping cream, whipped

(50g) dark chocolate, melted

(50g) whipping cream, add to dark chocolate

Dual Italian Cheesecake

雙色義大利起士蛋糕

做法 Directions

派底

1. 將馬斯卡邦起士和蛋黃混合；用滾水將吉利丁粉溶解；將動物性鮮奶油和植物性鮮奶油混合；將黑巧克力和動物性鮮奶油混合，用熱水隔水加熱。
2. 將馬斯卡邦起士蛋混合物和已溶化的吉利丁粉攪拌均勻，加入鮮奶油，攪拌滑順。
3. 將餡料分成2份，其中1份加入黑巧克力漿，攪拌滑順。

組合

1. 將一片巧克力蛋糕放於蛋糕模內，倒入巧克力餡料。
2. 放一片海綿蛋糕，再倒進原味餡料，置放冰箱冰1小時凝固後取出。
3. 在蛋糕面上放黑白巧克力碎片裝飾。

Filling

1. Combine mascarpone cheese with egg yolk. Dissolve gelatin powder in boiling water. Combine non-dairy cream with whipping cream. Combine dark chocolate with whipping cream, warming up in hot water.
2. Combine mascarpone cheese mixture with gelatin solution. Add cream. Blend until smooth.
3. Divide Filling into 2 portions. Add 1 portion to melted dark chocolate. Blend until smooth.

Assembly

1. Put a chocolate cake slice in a cake mould. Pour in chocolate filing.
2. Put a base cake slice on top. Pour in plain Filling. Refrigerate for 1 hour until it is firm. Take out.
3. Garnish with grated dark and white chocolate.

Q：為什麼不全用植物性鮮奶油或動物性鮮奶油，卻兩者同時用呢？
Why did you combine whipping cream and non-dairy cream together instead of either one?

A：動物性鮮奶油的質感細軟滑順，容易回軟；植物性鮮奶油就比較堅挺而耐放，兩者結合為一，互補不足，軟硬適中，適合初學者使用。
Whipping cream is so soft and smooth that it may collapse. Non-dairy cream is firmer. They are complementary to each other. The match is a good idea for beginners to follow.

材料 Ingredients

餅底

曲奇派底{1份}
(請參閱第122頁)

餡料

奶油乳酪{250克}
雞蛋{2個}
吉利丁粉{10克}
滾水{70克}
糖{10克}
植物性鮮奶油
{150克，已打發}
動物性鮮奶油
{100克，已打發}
白巧克力{120克}
牛奶{120克，白巧克力用}

裝飾

白巧克力{適量}
新鮮櫻桃{2粒}

Base

(1) cookie base
(refer to p.122)

Filling

(250g) cream cheese
(2) eggs
(10g) gelatin powder
(70g) boiling water
(10g) sugar
(150g) non-dairy cream, whipped
(100g) whipping cream, whipped
(120g) white chocolate
(120g) milk, add to white chocolate

Decoration

White chocolate
(2) fresh cherries

White Chocolate Cheesecake

白巧克力起士蛋糕

做法 Directions

餡料

1. 將奶油乳酪和雞蛋混合。
2. 用滾水將吉利丁粉煮溶。
3. 煮熱牛奶，加入白巧克力，煮至溶化。
4. 將起士糊和已溶化的吉利丁粉攪拌滑順，再加入牛奶巧克力攪拌均勻，慢慢拌入已打發的混合鮮奶油，攪拌滑順。

組合

1. 在模裏放入曲奇派底，輕輕壓實，倒進餡料，放入冰箱冰1小時。
2. 取出起士蛋糕，並在蛋糕面放白巧克力和新鮮櫻桃裝飾。

Filling

1. Combine cream cheese with egg.
2. Dissolve gelatin powder in boiling water.
3. Warm up milk. Add white chocolate. Cook until melted.
4. Blend cream cheese mixture and gelatin solution until smooth. Combine with melted milk chocolate. Gradually fold in mixed cream. Blend until smooth.

Assembly

1. Put cookie curst in a mould. Press gently with fingertips. Pouring in Filling. Refrigerate for 1 hour until it is firm.
2. Take out the cheesecake. Garnish top with white chocolate and fresh cherries.

Q：白巧克力牛奶煮好後，是否需要放涼後才拌入已打發的鮮奶油？
Having white chocolate melted in milk, should I wait until it is cool before fold in whipped cream?

A：是啊！因為白巧克力牛奶太熱會把鮮奶油溶掉，所以應該把它放涼或隔水置於冷水中，使白巧克力牛奶溫度降低後再與鮮奶油融合。
Absolutely! It is because cream melts if white chocolate is overly hot. Hence, you should let it cool or cool it down in water before you combine it with cream.

材料 Ingredients

派底

糖酥皮{1份}
(參閱第122頁)

巧克力蛋糕{2片}
(請參閱第121頁)

餡料

奶油乳酪{100克}

牛奶{100克}

C.P.卡士達粉{20克}

植物性鮮奶油{100克}

黑巧克力{50克}

動物性鮮奶油
{50克，黑巧克力用}

裝飾

巧克力碎片{適量}

Base

(1) sweet pastry
(Refer to p.122)

(2) slices chocolate cake
(Refer to p.121)

Filling

(100g) cream cheese

(100g) milk

(20g) custard powder

(100g) non-dairy cream

(50g) dark chocolate

(50g) whipping cream,
add to dark chocolate

Decoration

Chocolate, grated

Black Forest Cheese Pie

黑森林起士派

做法 Directions

糖鬆底

將糖酥皮放進派模裏，輕輕壓實，放入已預熱的烤箱，以200℃烤熟需時約15分鐘，再放涼。

餡料

1. 將奶油乳酪、牛奶和卡士達粉混合，攪拌滑順。
2. 將黑巧克力和動物性鮮奶油混合，隔水加熱。
3. 將黑巧克力漿拌入已混合的鮮奶油，慢慢攪拌滑順成巧克力鮮奶油。

組合

1. 在派皮上塗上一層巧克力溶液，待冷卻變乾。
2. 再擠上奶油乳酪餡和巧克力鮮奶油，鋪上巧克力碎片即可。

Sweet pastry

Put sweet short pastry into a pie mould. Gently press with fingertips. Bake in a preheated oven at 200°C for 15 minutes. Let cool.

Filling

1. Combine cream cheese with milk and custard powder. Blend until smooth.
2. Combine dark chocolate with whipping cream, warming up in hot water.
3. In the bowl of cream cheese custard, gradually fold in melted dark chocolate. Gently blend until it forms to chocolate cream.

Assembly

1. Brush the surface of pie curst with a layer of melted chocolate, Let cool and firm.
2. Pipe cream cheese and chocolate cream mixture. Garish with grated chocolate. Done.

Q：黑森林是德國著名蛋糕，因為什麼理由令你把蛋糕改為派呢？
Black forest is a famous Germany style cake. Why did you convert it to a pie?

A：黑森林是非常著名的蛋糕，但吃多了也會膩 ，我認為巧克力與脆派底的口感也很匹配，所以改掉傳統，變為派皮形式，很受學生歡迎，才引入這書裏。
I like to be creative. Chocolate quite matches the texture of pie crust. This innovative recipe is popular among my students.

Q：為甚麼要用酸奶油做派面裝飾？
Why is sour cream used for decoration?

A：這是典型美式起士蛋糕的組合，因為它含濃郁起士味，質感充實，為了減輕膩口的感覺，以及引發風味，所以會利用酸奶油平衡味道。
It is a traditional American cheesecake recipe. It is rich and firm. Sour cream may balance out the heaviness.

難度●●●中

American Cheese Tart
美式起士塔

材料 Ingredients

餅底
糖酥皮{1份}
(參閱第122頁)

餡料
奶油乳酪{250克}
糖{50克}
雞蛋{3個}
奶油溶液{30克}

裝飾
酸奶油{100克}
糖粉{50克}

Base
(1) sweet short pastry
(Refer to p.122)

Filling
(250g) cream cheese
(50g) sugar
(3) eggs
(30g) butter, melted

Decoration
(100g) sour cream
(50g) icing sugar

做法 Directions

塔底

將糖酥皮放進塔模裏，輕輕壓實，放入已預熱的烤箱，以200℃烤熟需時約15分鐘。

餡料

1. 將奶油乳酪放軟，加入糖攪拌滑順，再放入雞蛋攪拌均勻至形態像鮮奶油一般。
2. 最後把奶油溶液混合，備用。

組合

1. 取出塔底，倒進餡料，放入已預熱的烤箱，用180℃烤15-20分鐘。
2. 把酸奶油與糖粉拌至細滑，放在起士塔上，抹勻，放回烤箱烤5分鐘即可。

Pie crust

Put sweet short pastry into a pie mould. Gently press with fingertips. Bake in a preheated oven at 200°C for 15 minutes. Let cool.

Filling

1. Leave cream cheese at room temperature until softened. Add sugar. Blend until smooth. Add egg. Blend until it is creamy.
2. Combine with melted butter. Set aside.

Assembly

1. Take out tart crust. Pour in filling. Bake in a preheated oven at 180°C for 15-20 minutes.
2. Blend sour cream and icing sugar to smooth. Spread over top evenly. Bake in an oven for 5 minutes. Done.

材料 Ingredients

派底

鬆酥{1份}(參閱第122頁)
糖粉{適量，灑表面用}

餡料

奶油乳酪{250克，放軟}
糖{40克}，雞蛋{2個}
蘋果{2個，切片}
糖{30克，煮成焦糖}

奶酥

奶油{30克}
糖{30克}
麵粉{60克}
肉桂粉{1克}

Base

(1) crisp crust pastry
(Refer to p.122)

Icing sugar, sprinkle over the surface

Filling

(250g) cream cheese, softened

(40g) sugar

(2) eggs

(2) apple, sliced

(30g) sugar, sprinkle over apple

Crumble

(30g) butter

(30g) sugar

(60g) flour

(1g) cinnamon powder

Apple Crumble Cheesecake
金寶蘋果起士蛋糕

做法 Directions

派底

將派皮做成直徑8吋的大小，鋪進模裏，輕輕按實，放進烤箱，用200℃烤10分鐘。

餡料

1. 將奶油乳酪和糖混合，一起攪拌滑順，加入雞蛋，再次攪拌滑順。
2. 蘋果去皮、切粒；將糖煮至金黃色，加入蘋果，煮成焦糖蘋果。
3. 將奶油乳酪和焦糖蘋果混合，做成餡料。

奶酥

將製作奶酥材料全部搓勻，並搓成粗粒備用。

組合

1. 取出派底，放進餡料，餅面放奶酥，放入烤箱，用180℃烤20-25分鐘即可。
2. 取出後放片刻，灑上糖粉。

Pie crust

Put crisp crust pastry into a 8 inch mould. Gently press with fingertips. Bake in a preheated oven at 200°C for 10 minutes. Let cool.

Filling

1. Combine cream cheese with sugar. Blend until smooth. Add egg. Blend until smooth.
2. Peel apple and cut in cubes. Cook sugar until golden. Add apple and it forms to caramelized apple.
3. Combine cream cheese with caramelized apple and it forms to filing.

Crumble

Rub all the ingredients of crumble until it from crumbs. Set aside.

Assembly

1. Take out tart crust. Pour in filling. Top with crumble. Bake at 180°C for 20-25 minutes. Done.
2. Take out. Let cool for a brief while. Sprinkle with icing sugar.

Q：奶酥源自何處？
Where did crumble originate?

A：奶酥是源自第二次世界大戰的英國愛爾蘭，主要用奶油、麵粉和糖揉為粗粒，撒在甜點和派皮，除了裝飾外，還會有甜脆口感，散發奶油濃香，美國甜點廣泛使用。
It was created in Ireland during the Second World War. The main ingredients are butter, flour and sugar. Rub them into crumble and place top of dessert and pie. Other than for decoration, it enhances the crunchiness and buttery flavour. Crumble is widely used in American dessert recipes.

材料 Ingredients

餅底

海綿蛋糕｛2片｝
（參閱第121頁）

餡料

茅屋起士｛1盒｝
糖｛40克，茅屋起士用｝
火龍果汁｛100克｝
牛奶｛50克｝
植物性鮮奶油｛150克，已打發｝
動物性鮮奶油｛100克，已打發｝
吉利丁粉｛12克｝
滾水｛70克｝
糖｛10克，吉利丁粉用｝

裝飾

火龍果片｛適量｝
巧克力裝飾｛適量｝
植物性鮮奶油｛100克，已打發｝

Base

(2) slices base cake
(Refer to p.121)

Filling

(1) box cottage cheese
(40g) sugar, for cottage cheese
(100g) dragon juice
(50g) milk
(150g) non-dairy cream, whipped
(100g) whipping cream, whipped
(12g) gelatin powder
(70g) boiling water
(10g) sugar, for gelatin powder

Decoration

Dragon fruit slices
Chocolate
(100g) non-dairy cream, whipped

難度●●●中

Cottage Cheese and Dragon Fruit Cheesecake

茅屋起士火龍果蛋糕

做法 Directions

餡料

1. 將茅屋起士和糖攪拌滑順。
2. 將火龍果汁和牛奶攪拌均勻。
3. 用滾水把吉利丁粉和糖拌至溶解。
4. 將已溶化的吉利丁粉和茅屋起士糊混合，攪拌滑順，加入牛奶火龍果汁攪拌均勻後，再慢慢拌入已打發的鮮奶油，攪拌均勻。

組合

1. 在糕模內放進一片海綿蛋糕，倒入半份餡料，再放另一片海綿蛋糕，然後放於冰箱冰2小時使蛋糕凝固。
2. 取出起士蛋糕，抹一層薄鮮奶油，並用火龍果片和巧克力裝飾。

Filling

1. Combine cottage cheese with sugar.
2. Combine dragon fruit juice with milk.
3. Dissolve gelatin powder and sugar in boiling water.
4. Combine gelatin solution with cottage cheese mixture. Ass dragon fruit milk. Blend. Gradually fold in whipped cream. Combine.

Assembly

1. Put a base cake slice in a cake mould. Pour in half of Filling. Place another base cake slice on top. Refrigerate for 2 hours until it is firm.
2. Take out cheesecake. Spread over top a thin layer of cream. Garnish with dragon fruit slices and chocolate.

Q：為甚麼要選火龍果做起士蛋糕？
Why is dragon fruit you choice?

A：因為火龍果清甜多汁，味道輕盈，質感很好，果肉有白、粉紅和黃色，容易與甜點配色和配味。它原產於墨西哥、加勒比海和中美洲，之後引進中國、台灣、馬來西亞和越南等地栽培，屬熱帶和亞熱帶的優質水果。
It is sweet, light and juicy. Plus its flesh comes in white, pink and yellow colour. Dragon fruits grow in many countries. It may go well with cheesecake.

材料 Ingredients

蛋糕底

海綿蛋糕{ 2片 }
（請參閱第121頁）

餡料

低脂起士{ 150克 }
馬斯卡邦起士{ 100克 }
蛋黃{ 2個 }
糖{ 20克，調起士糊用 }
吉利丁粉{ 12克 }
滾水{ 70克 }

咖啡糖水

咖啡{ 10克 }
糖{ 20克 }
滾水{ 50克 }

義式蛋白霜

蛋白{ 150克 }
糖{ 70克 }
滾水{ 50克 }

裝飾

可可粉{ 適量 }

Base

(1) slice base cake
(Refer to p.121)

Filling

(150g) low fat cream cheese
(100g) mascarpone cheese
(2) egg yolks
(20g) sugar, for cheese batter
(12g) gelatin powder
(70g) boiling water

Coffee syrup

(10g) coffee
(20g) sugar
(50g) boiling water

Italianstyle meringue

(150g) egg white
(70g) sugar
(50g) boiling water

Decoration

Cocoa powder

Low Fat Italian Cheesecake

低脂義式起士蛋糕

做法 Directions

義式蛋白霜

用滾水與糖煮至120℃，倒入已打發的蛋白。

咖啡糖水

將咖啡、糖和滾水混合。

餡料

1. 將低脂起士和馬斯卡邦起士攪拌滑順。
2. 將蛋黃和糖打發；吉利丁粉和滾水拌至溶化。
3. 將蛋黃糊和已溶化的吉利丁粉混合，攪拌滑順，加入起士糊，攪拌均勻。

組合

1. 在海綿蛋糕片，灑上咖啡糖水，放入糕模中，倒入半份餡料，再放入另一片海綿蛋糕，放入冰箱冰2小時，使蛋糕凝固。
2. 取出蛋糕，灑上可可粉。

Italia style meringue

Cook sugar in boling water at 120°C. Fold in beaten egg white. It forms Italian style meringue.

Coffee syrup

To make coffee syrup, dissolve coffee and sugar in boiling water.

Filling

1. Blend low fat cream cheese and mascarpone cheese together to smooth.
2. Beat egg yolk and sugar. Dissolve gelatin powder in boiling water.
3. Combine beaten egg yolk with gelatin solution. Blend until smooth. Add cheese mixture. Combine.

Assembly

1. Brush a base cake slice with coffee syrup. Put it into a cake mould. Pour in half of Filling. Place another base cake slice on top. Pour in the remainder of Filling. Refrigerate for 2 hours.
2. Take out the cheesecake. Sprinkle over top with cocoa powder.

Q：用義式蛋白霜做這蛋糕有何用處？
Why did you use Italian style meringue?

A：義式蛋白霜因用了煮糖的方法調配，糖漿經加熱煮稠，濃度會高一點且帶有韌度，使蛋白霜有點力度，不易變回液態。
Syrup turns to thicken over heat and it increases its strength so that meringue will not collapse easily.

材料 Ingredients

蛋糕

蛋白{ 3個 }
蛋黃{ 3個 }
麵粉{ 50克 }
發粉{ 1克 }
糖{ 40克 }
牛奶{ 50克 }
玉米油{ 20克 }

低脂起士餡

動物性鮮奶油{ 250克，打發 }
糖粉{ 30克 }
低脂起士{ 150克 }
吉利丁粉{ 6克 }
清水{ 25克 }

裝飾

新鮮覆盆子{ 100克 }

Cake

(3) egg white
(3) egg yolks
(50g) flour
(1g) baking powder
(40g) sugar
(50g) milk
(20g) corn oil

Low fat cheese fillings

(250g) whipping cream, whipped
(30g) icing sugar
(150g) low fat cheese
(6g) gelatine powder
(25g) water

Decoration

(100g) fresh raspberries

難度●●●中

Low Fat Angle Cheesecake

天使低脂起士蛋糕

做法 Directions

蛋糕

1. 蛋白打發，加入糖打至濃稠。
2. 把蛋黃、牛奶、麵粉、發粉和玉米油混合攪拌至細滑。
3. 分3次拌入已打發的蛋白拌至細滑，倒入餅模。
4. 烤箱預熱至180℃，放進烤箱烘烤20分鐘後，取出，放涼。

低脂起士餡

1. 吉利丁粉和清水拌至溶解。
2. 將已打發動物性鮮奶油和糖粉拌至細滑，再拌入低脂起士，然後加入已溶化的吉利丁粉攪拌均勻。

組合

將已冷凍的蛋糕，抹上低脂起士餡，再綴上新鮮覆盆子裝飾。

Cake

1. Beat egg white until stiff. Add sugar and beat to thicken.
2. Mix egg yolks, milk, flour, ba ur into cake mould.
3. Preheat an oven at 180ºC. Transfer the cake in the oven and bake for 20 minutes. Take out. Let cool.

Low fat cheese fillings

1. Dissolve gelatine powder in water.
2. Beat whipped cream and icing sugar to smooth. Fold in low fat cheese. Then add the gelatine solution.

Assembly

Spread the low fat cheese filling onto the cold cake. Decorate with raspberries.

Q：天使蛋糕為甚麼會用中空圓模呢？
Why is a tube up at the centre of angel cake pan?

A：天使蛋糕的特質是輕巧和鬆泡泡，入口如風。除了用蛋白和換了玉米粉外，還因蛋糕模的中空圓柱體設計，能在烘焙過程中均勻地全面受熱，使它的大部份表面積緊貼蛋糕模膨脹，加上環形造型可讓蛋糕快速冷卻，而有綿軟鬆化口感。
Texture of angel cake is light and puffy, like a breeze passing in your mouth. The tube helps to spread heat evenly and cake expands better. The design also speed up cooling to keep softness.

材料 Ingredients

餅底

海綿蛋糕{2片}
(參閱第121頁)

餡料

奶油乳酪{250克}
糖{20克，調起士漿用}
青蘋果汁{100克}
蘋果優格{100克}
白酒{20克}
吉利丁粉{14克}
糖{10克}
滾水{70克}
植物性鮮奶油{150克，已打發}
動物性鮮奶油{100克，已打發}

裝飾

水果粒{適量}
植物性鮮奶油
{100克，已打發}
食用色素
{少許，調鮮奶油色澤}

Base

(2) slice base cake
(Refer to p.121)

Filling

(250g) cream cheese
(20g) sugar, for cream cheese
(100g) green apple juice
(100g) apple yogurt
(20g) white wine
(14g) gelatin powder
(10g) sugar
(70g) boiling water
(150g) non-dairy cream, whipped
(100g) whipping cream, whipped

Decoration

Fruit, cubed
(100g) non-dairy cream, whipped
Colouring, for cream

難度●●●高

Green Apple Mousse Cheesecake

青蘋果慕斯起士蛋糕

做法 Directions

餡料

1. 將奶油乳酪、糖、青蘋果汁、蘋果優格和白酒混合，攪拌滑順。
2. 用滾水把吉利丁粉和糖混合溶解。
3. 將奶油乳酪蘋果漿和已溶化的吉利丁粉混合，攪拌滑順，加入已打發的鮮奶油攪拌均勻。

組合

1. 在模型中放入一片海綿蛋糕，倒入半份餡料，再放另一片海綿蛋糕，再次倒入其餘餡料，放入冰箱冰2小時使蛋糕凝固。
2. 取出起士蛋糕，把已與食用色素混合的植物性鮮奶油攪拌均勻，再用水果和巧克力片裝飾。

Filling

1. Combine cream cheese with sugar, green apple juice, apple yogurt and white wine. Blend until smooth.
2. Dissolve gelatin powder and sugar in water.
3. Combine apple cream cheese mixture with gelatin solution. Blend until smooth add whipped cream. Combine.

Assembly

1. Put a base cake slice into a cake mould. Pour in half of Filling. Place another base cake slice on top. Pour in the remainder of Filling. Refrigerate for 2 hours.
2. Take out the cheesecake. Combine non-dairy cream with colouring. Spread over top. Garnish with fruits and chocolate flakes.

Q：是否必須用白酒調味？
Is it nece white wine?

A：在甜點裏添加酒，可提升奶油乳酪的香味，要是給小朋友吃就不加也可以。
It enhances the aroma of cheese. You may skip it if you serve kids.

Q：柚子醬是現代甜點的新寵兒？
Is it a new trend to use pomelo jam?

A：沒錯，柚子醬是韓國特產，採用皮厚金黃的柚子，調配蜜糖做成如果醬似的製品，可以沖水、做菜和甜點，有很濃烈的果香味道，因有濃重的橘油香的微苦回甘，與蜜糖的甜很平衡。
Yup, pomelo jam is a Korean product. You may make tea, cook meals or make dessert with pomelo jam. You may smell the mellow, fruity aroma. It slightly contains bitterness, which balances the sweetness of honey.

難度●●●高

Pomelo Cheesecake
柚子起士蛋糕

材料 Ingredients

餅底
海綿蛋糕{2片}
(請參閱第121頁)

餡料
奶油乳酪{250克，放軟}
柚子醬{70克}
牛奶{80克}
蛋黃{2個}
糖{20克，調奶油乳酪漿用}
吉利丁粉{12克}
糖{10克}
滾水{70克}
植物性鮮奶油{150克，已打發}
動物性鮮奶油{100克，已打發}

裝飾
新鮮柚子果肉{適量}
巧克力{適量}
新鮮水果{適量}

Base
(2) slices base cake
(Refer to p.121)

Filling
(250g) cream cheese, softened
(70g) pomelo jam
(80g) milk
(2) egg yolks
(20g) sugar, for cream cheese
(12g) gelatin powder
(10g) sugar
(70g) boiling water
(150g) non-dairy cream, whipped
(100g) whipping cream, whipped

Decoration
Fresh pomelo pulps
Chocolate
Fresh fruits

做法 Directions

餡料
1. 將奶油乳酪、柚子醬和牛奶攪拌均勻。
2. 用滾水將吉利丁粉和10克糖溶解；將蛋黃和20克糖打發。
3. 將已溶化的吉利丁粉和蛋黃漿攪拌均勻，加入柚子牛奶漿，攪拌滑順。
4. 最後拌入已打發的鮮奶油，攪拌滑順。

組合
1. 在模裏放入一片海綿蛋糕，倒進半份餡料，再另放一片海綿蛋糕，置放入冰箱冰2小時使蛋糕凝固。
2. 取出起士蛋糕，用新鮮柚子果肉、巧克力和新鮮水果裝飾。

Filling
1. Blend cream cheese, pomelo jam and milk until smooth.
2. Dissolve gelatin powder and 10g sugar in boiling water. Beat egg yolk and 20g sugar.
3. Blend gelatin solution and beaten egg yolk until smooth. Add pomelo milk. Blend until smooth.
4. Fold in whipped cream. Blend until smooth.

Assembly
1. Put a base cake slice into a cake mould. Pour in half of Filling. Place another base cake slice on top. Pour in the remainder of Filling. Refrigerate for 2 hours.
2. Take out the cheese cake. Garnish with pomelo pulps, chocolate and fresh fruits.

材料 Ingredients

餅底

海綿蛋糕{2片}
(請參閱第121頁)

餡料

奶油乳酪{250克}
糖{20克，調起士漿用}
蛋黃{2個}
糖{20克，蛋黃用}
吉利丁粉{12克}
糖{10克，吉利丁粉用}
滾水{70克}
植物性鮮奶油{150克，已打發}
動物性鮮奶油{100克，已打發}
百香果果汁{80克}
牛奶{100克}

裝飾

芒果果凍粉{50克}
滾水{100克}

Base

(2) slices base cake
(Refer to p.121)

Filling

(250g) cream cheese
(20g) sugar, for cream cheese
(2) egg yolks
(20g) sugar, for egg yolk
(12g) gelatin powder
(10g) sugar, for gelatin powder
(70g) boiling water
(150g) non-dairy cream, whipped
(100g) whipping cream, whipped
(80g) passion fruit juice
(100g) milk

Decoration

(50g) mango jelly powder
(100g) boiling water

難度●●●高

Passion Fruit Mousse Cheesecake

百香果慕斯起士蛋糕

做法 Directions

裝飾

把所有材料攪拌均勻。

餡料

1. 將奶油乳酪和糖打發。
2. 將蛋黃和糖隔水加熱打發；用滾水將吉利丁粉和糖溶解。
3. 煮滾牛奶，加入蛋漿攪拌均勻，再加入已溶化的吉利丁粉攪拌均勻，拌入奶油乳酪糖漿至融合。
4. 拌入已打發的鮮奶油，攪拌滑順，最後倒進百香果果汁快速攪拌滑順。

組合

1. 在模裏放入一片海綿蛋糕，倒進半份餡料，再放另一片海綿蛋糕，倒進剩餘餡料，放入冰箱冰2小時使蛋糕凝固。
2. 取出起士蛋糕，倒進芒果果凍水裝飾後，放回冰箱冰。

Decoration

Dissolve mango jelly power in boiling water. Set aside.

Filling

1. Blend cream cheese and sugar until smooth.
2. In a bowl, beat egg yolk and sugar in hot water. Dissolve gelatin powder and sugar in boiling water.
3. Bring milk to a boil. Add to beaten egg. Combine. Add gelatin solution. Blend until smooth. Pour in passion fruit juice. Blend fast until thoroughly combined.
4. Fold in whipped cream. Blend until smooth. Pour in passion fruit juice. Blend fast until smooth.

Assembly

1. Put a base cake slice into a cake mould. Pour in half of Filling. Place another base cake slice on top. Pour in the remainder of Filling. Refrigerate for 2 hours.
2. Take out the cheesecake. Pour in mango jelly solution. Refrigerate until it is firm.

Q：百香果果應該如何選擇？
How should I select passion fruits?

A：買新鮮百香果果可選成熟而香味濃，柔身一點會不錯。不過香港在6-8月時偶有出產，只是香味比外國進口淡一點。市面也有百香果果泥可用，品質比較穩定，要是遇到當季時，可把百香果果挖出果肉，放入保鮮盒冷凍備用。
Pick ripe and aromatic passion fruits. It is okay if it is tender. Local passion fruits are available between June and August. Their taste is plainer than those imported. Quality of passion fruit puree is more stable. You may freeze some extra portions of fresh passion fruit to store up.

材料 Ingredients

餅底

消化餅底{1份}
（請參閱第122頁）

海綿蛋糕{1片}
（請參閱第121頁）

餡料

馬斯卡邦起士{250克}

蛋黃{2個，攪拌滑順}

覆盆子汁{80克}

牛奶{80克}

吉利丁粉{12克}

滾水{70克}

糖{30克，溶解}

動物性鮮奶油{100克，已打發}

植物性鮮奶油{150克，已打發}

裝飾

覆盆子{適量}

植物性鮮奶油{100克，已打發}

覆盆子汁{20克}

Base

(1) digestive biscuit base
(Refer to p.122)

(1) slice base cake (Refer to p.121)

Filling

(250g) mascarpone cheese

(2) egg yolks, blend to smooth

(80g) raspberry juice

(80g) milk

(12g) gelatin powder

(70g) boiling water

(30g) sugar, dissolved

(100g) whipping cream, whipped

(150g) non-dairy cream, whipped

Decoration

Raspberries

(100g) non-dairy cream, whipped

(20g) raspberry juice

難度●●●高

Raspberry Mousse Cheesecake
覆盆子慕斯起士蛋糕

做法 Directions

餡料

1. 將馬斯卡邦起士和蛋黃混合，攪拌滑順。
2. 用滾水把吉利丁粉和糖拌至溶解。
3. 將起士蛋漿、已溶化的吉利丁粉、覆盆子汁和牛奶混合攪拌均勻，再與已打發的鮮奶油混合攪拌均勻。

組合

1. 在糕模裏放進消化餅底輕壓，倒進餡料，放入烤盤再倒入熱水，放入已預熱的烤箱。
2. 將起士蛋糕用160℃爐溫烤45分鐘，取出，放涼。
3. 把植物性鮮奶油與覆盆子汁攪拌均勻，抹在蛋糕上，再用覆盆子裝飾即可。

Filling

1. Combine mascarpone cheese with egg yolk. Blend until smooth.
2. Dissolve gelatin powder and sugar in boiling water.
3. Blend cheese batter, gelatin solution, raspberry juice and milk until smooth. Combine with whipped cream. Blend until smooth.

Assembly

1. Put digestive biscuit base in a cake mould. Gently press with fingertips. Pour in Filling. Put in a tray sheet filled with hot water. Place in a preheated oven.
2. Bake the cheese cake at 160°C for 45 minutes. Take out. Let cool.
3. Combine non-dairy cream with raspberry juice. Spread over top of the cake. Garnish with raspberries.

Q：做餡料用的鮮奶油，究竟應該打發至什麼程度才合標準？
For Filling, how stiff should the cream look like?

A：做餡料的鮮奶油不要過份打發或堅挺，最佳狀況是剛從液體狀變固體，仍然有少許微流動的狀況，這樣的質感會細滑，適合做餡料。
Do not blend overly or until too stiff. It is the best consistency, fine and smooth, when it just solidifies and is slightly fluidly.

Q：消化餅底壓在餅模時，不用壓得那麼實？
I should not press digestive biscuit base firm?

A：做起士蛋糕時，往往會用消化餅做底，千萬不要壓得那麼硬實，否則會很硬又難吃；但又不能壓得太鬆，否則很容易就散掉。
It is not a pleasant experience if someone eats firm digestive biscuit base. But if you don't press enough, it is hard to form the shape.

難度●●●高

Papaya Cheesecake

木瓜起士蛋糕

材料 Ingredients

餅底

消化餅底{1份}
(請參閱第122頁)
海綿蛋糕{1片}
(請參閱第121頁)

餡料

馬斯卡邦起士{250克}
萬壽果(即木瓜)汁{200克}
牛奶{100克}
吉利丁粉{14克}
糖{20克}
滾水{70克}
植物性鮮奶油{150克，已打發}
動物性鮮奶油{100克，已打發}

裝飾

木瓜{適量}
櫻桃{數粒}
巧克力條{2條}

Base

(1) digestive biscuit base
(Refer to p.122)

(1) slice base cake
(Refer to p.121)

Filling

(250g) mascarpone cheese

(200g) papaya juice

(100g) milk

(14g) gelatin powder

(20g) sugar

(70g) boiling water

(150g) non-dairy cream, whipped

(100g) whipping cream, whipped

Decoration

Papaya

Cherries

(2) chocolates sticks

做法 Directions

餡料

1. 將木瓜果汁和牛奶攪拌均勻。
2. 用滾水把吉利丁粉和糖溶解。
3. 將馬斯卡邦起士、木瓜牛奶和已溶化的吉利丁粉一同攪拌均勻，慢慢拌入已打發的鮮奶油攪拌均勻。

派底

1. 先在模裏放入消化餅底輕壓。
2. 倒入餡料，放入冰箱冰2小時使其凝固。
3. 取出凝固的餅乾，餅面放木瓜片、櫻桃和巧克力條裝飾。

Filling

1. Combine papaya juice with milk.
2. Dissolve gelatin powder and sugar in boiling water.
3. Blend mascarpone cheese, papaya milk and gelatin solution until smooth. Gradually fold in whipped cream. Combine.

Assembly

1. Put digestive biscuit base in a cake mould. Press gently with fingertips.
2. Pour in Filling. Refrigerate for 2 hours until it is firm.
3. Take out the cake. Garish top with papaya slices, cherries and chocolate.

材料 Ingredients

餅底

清蛋糕{ 2片 }
（參閱第121頁）

餡料

奶油乳酪{ 250克 }
糖{ 20克 }
榴槤{ 200克 }
牛奶{ 100克 }
吉利丁粉{ 12克 }
糖{ 20克 }
滾水{ 70克 }
植物性鮮奶油{ 250克，已打發 }

抹茶鮮奶油裝飾

植物性鮮奶油{ 200克，已打發 }
抹茶粉{ 1茶匙 }
滾水{ 2湯匙 }

Base

(1) slices base cake
(Refer to p.121)

Filling

(250g) cream cheese
(20g) sugar
(200g) durian
(100g) milk
(12g) gelatin powder
(20g) sugar
(70g) boiling water
(250g) non-dairy cream, whipped

Matcha cream Decoration

(200g) non-dairy cream, whipped
(1) tsp matcha powder
(2) tbsp boiling water

難度●●●高

Durian Mousse Cheesecake

榴槤慕斯起士蛋糕

做法 Directions

抹茶鮮奶油裝飾

抹茶粉與滾水調勻，拌入已打發的鮮奶油。

餡料

1. 打發奶油乳酪和糖。
2. 將榴槤和牛奶攪拌成果泥狀。
3. 用滾水把將吉利丁粉和糖攪拌至溶解。
4. 將奶油乳酪糖漿放入已溶化的吉利丁粉中，攪拌均勻，加入榴槤果泥奶攪拌均勻，慢慢拌入植物性鮮奶油，攪拌均勻。

組合

1. 將一片海綿蛋糕片放入模中，倒入半份餡料，再放另一片海綿蛋糕，倒入剩餘餡料，放入冰箱冰2小時使蛋糕凝固。
2. 取出榴槤起士蛋糕，用綠茶鮮奶油擠花裝飾。

Matcha cream Decoration

Combine matcha powder with boiling water. Fold in whipped cream.

Filling

1. Blend cream cheese and sugar until smooth.
2. Blend durian and milk in a blender until it forms to puree.
3. Dissolve gelatin powder and sugar in boiling water.
4. Combine cream cheese syrup with gelatin solution. Add durian milk puree. Combine. Gradually fold in non-dairy cream. Combine.

Assembly

1. Put a base cake slice into a cake mould. Pour in half of Filling. Place another base cake slice on top. Pour in the remainder of Filling. Refrigerate for 2 hours.
2. Take out the durian cheesecake. Spread over top with matcha cream. Pipe for Decoration.

Q：做抹茶鮮奶油，是否任何抹茶粉均可？
Can I use matcha powder of any kind?

A：任何綠茶粉都可以，依自己喜愛品牌和味道就可以了。
Yes, it's up to your preference.

材料 Ingredients

餅底

海綿蛋糕{1片}
(參閱第121頁)

餡料

馬斯卡邦起士{250克}
蛋黃{2個}
糖{40克}
牛奶{100克}
吉利丁粉{12克}
糖{10克}
滾水{70克}
植物性鮮奶油{150克，已打發}
動物性鮮奶油{100克，已打發}
綠茶粉{6克}
滾水{1湯匙}
日本紅豆罐頭{50克}

裝飾

吉利丁粉{6克}
糖{5克}
滾水{35克}
日本紅豆罐頭{25克}

Base

(2) slices base cake
(Refer to p.121)

Filling

(250g) mascarpone cheese
(2) egg yolks
(40g) sugar
(100g) milk
(12g) gelatin powder
(10g) sugar
(70g) boiling water
(150g) non-dairy cream, whipped
(100g) whipping cream, whipped
(6g) mat cha powder
(1) tbsp boiling water
(50g) canned red beans

Decoration

(6g) gelatin powder
(5g) sugar
(35g) boiling water
(25g) canned red beans

難度●●●高

Red Bean and Green Tea Cake
紅豆綠茶慕斯蛋糕

做法 Directions

餡料

1. 將蛋黃和糖打發；將牛奶煮滾，加入蛋漿中攪拌滑順。
2. 用滾水將吉利丁粉和糖溶解。
3. 將馬斯卡邦起士、蛋漿和已溶化的吉利丁粉混合攪拌均勻，拌入已打發的鮮奶油，分成兩份。
4. 綠茶粉用1湯匙水調勻，與半份起士漿拌好；另半份起士漿加入日本紅豆，攪拌均勻。

組合

1. 在模放入1 片海綿蛋糕，再倒入紅豆餡料，放入冰箱，冰1小時。
2. 取出蛋糕，倒入綠茶起士餡，放入冰箱冰1小時直到餡料凝固。
3. 把裝飾用的吉利丁粉、糖和滾水攪拌均勻，倒在紅豆綠茶起士蛋糕上，放上日本紅豆，再淋上已溶化的吉利丁粉，放入冰箱直至蛋糕凝固即可。

小提醒

1. 綠茶粉要與滾水調開才容易與起士漿混合。
2. 起士漿應分兩份各自做，否則在等待的期間，另一份餡料已冷了。

Filling

1. Beat egg yolk and sugar. Bring milk to a boil and then combine with beaten egg yolk. Blend until smooth.
2. Dissolve gelatin powder in boiling water.
3. Blend mascarpone cheese, beaten egg and gelatin solution until smooth. Fold in whipped cream. Divide in 2 portions.
4. Combine matcha powder with 1 tbsp water. Combine with half of the cheese batter. Combine another half of cheese batter with red beans.

Assembly

1. Put a base cake slice in a mould. Pour in red bean Filling. Refrigerate for 1 hour until it is firm.
2. Take out the cake. Pour in green tea cheese Filling. Refrigerate for another hour until it is firm.
3. To decorate, dissolve gelatin powder and sugar in boiling water. Spread over top of the cheese cake. Put red beans. Sprinkle with gelatin solution. Refrigerate until it is firm.

Tip

1. It is easy to combine with cheese mixture after matcha powder has dissolved in boiling water.
2. Prepare two portions of cheese mixture respectively. Otherwise, one portion may turn hard while the wait.

Q：做雙色起士蛋糕，是否需要逐層製作？
Should I prepare each layer at a time?

A：當然，因為做雙色起士蛋糕，要有耐性，一層一層做須待一層凝固才可以做另一層，否則兩者因未凝固而變混濁，層次不分明。
Sure. It takes the time to make layers. You have to be patient that layers must be firm before you pour in another portion. Otherwise, it will mess up.

材料 Ingredients

餅底

鬆酥餅底{1份}
（請參閱第122頁）

巧克力溶液
{50克，灑餅底用}

海綿蛋糕{1片}
（請參閱第121頁）

餡料

馬斯卡邦起士{250克}

南瓜泥{250克}

牛奶{80克}

吉利丁粉{14克}

糖{40克}

滾水{70克}

植物性鮮奶油{150克，已打發}

動物性鮮奶油{100克，已打發}

裝飾

熟南瓜{1片}

Base

(1) shortcrust pastry(Refer to p.122)

(50g) melted chocolate, brush on base

(1) slice base cake (Refer to p.121)

Filling

(250g) mascarpone cheese

(250g) mashed pumpkin

(80g) milk

(14g) gelatin powder

(40g) sugar

(70g) boiling water

(150g) non-dairy cream, whipped

(100g) whipping cream, whipped

Decoration

(1) sliced cooked pumpkin

難度●●●高

Pumpkin Cheesecake

南瓜起士蛋糕

做法 Directions

派皮

將派皮做成直徑8吋，鋪進模裏，輕輕按實，放進烤箱，用200℃烤15分鐘。

餡料

1. 用滾水把吉利丁粉和糖拌至溶解。
2. 將馬斯卡邦起士、南瓜泥和牛奶混合，攪拌均勻。
3. 放入已溶化的吉利丁粉攪拌均勻，再輕輕拌入已打發的鮮奶油攪拌滑順。

組合

1. 把巧克力溶液刷在派底上，待乾，放一片海綿蛋糕，倒入起士餡抹平，放進冰箱冰2小時。
2. 取出起士蛋糕，餅面放一片熟南瓜裝飾。

Pie crust

To make crust: Cut pastry into the diameter of 8 inches. Place in a mould. Press gently with fingertips. Bake at 200°C for 15 minutes.

Filling

1. Dissolve gelatin powder and sugar in boiling water.
2. Thoroughly combine mascarpone cheese with mashed pumpkin and milk.
3. Pour in gelatin solution. Combine. Gently fold in whipped cream. Blend until smooth.

Assembly

1. Brush pie crust with melted chocolate. Let dry. Put a base cake slice. Pour in cheese Filling. Level. Refrigerate for 2 hours until it is firm.
2. Take out the cheesecake. Garnish with a cooked pumpkin slice.

Q：南瓜泥可以自己做，還是買罐裝好？
Should use fresh pumpkin puree or canned pumpkin?

A：當然自己做的比較好！因為當季的南瓜品種多，價錢實惠，味道天然又健康，挑選好品種，夠柔軟就好，蒸熟壓泥即可食用。若不是當季，就可買罐裝！
Fresh pumpkin puree of course! A variety of natural, healthy pumpkins are available in the market during the harvest season. Pick tender pumpkins. Steam and mash--there you go!

材料 Ingredients

胡蘿蔔蛋糕
- 麵粉{ 70克 }
- 全麥粉{ 35克 }
- 發粉{ 2克 }
- 蘇打粉{ 2克 }
- 肉桂粉{ 1克 }
- 奶油{ 60克 }
- 糖{ 70克 }
- 雞蛋{ 1個 }
- 胡蘿蔔絲{ 100克 }
- 葡萄乾{ 20克 }
- 核桃碎粒{ 20克 }

餡料
- 奶油乳酪{ 100克 }
- 優格{ 50克 }
- 吉利丁粉{ 4克 }
- 糖{ 20克 }
- 滾水{ 20克 }
- 檸檬{ 半個，榨汁 }

裝飾
- 杏仁糕{ 20-30克 }
- 橙色食用色素{ 少許，做胡蘿蔔 }
- 綠色食用色素{ 少許，做胡蘿蔔梗 }
- 胡蘿蔔絲{ 適量 }

Carrot cake
- (70g) flour
- (35g) whole wheat flour
- (2g) baking powder
- (2g) baking soda
- (1g) cinnamon powder
- (60g) butter
- (70g) sugar
- (1) egg
- (100g) carrot, shredded
- (20g) raisins
- (20g) walnuts, crushed

Filling
- (100g) cream cheese
- (50g) yogurt
- (4g) gelatin powder
- (20g) sugar
- (20g) boiling water
- (½) lemon, squeezed

Decoration
- (20-30g) almond paste
- Orange colouring, for imitating carrot
- Green colouring, for imitating stem
- Shredded carrot

難度●●●高

Carrot Yogurt Cheesecake
胡蘿蔔優格起士蛋糕

做法 Directions

胡蘿蔔蛋糕

1. 把全部粉材料過篩，備用。
2. 先將奶油和糖打至鬆軟變白，加入雞蛋攪勻，再拌入粉性材料，最後加入剩餘材料攪拌均勻。
3. 倒入已墊烘焙紙的蛋糕模裏，用180℃烤20分鐘，取出蛋糕，放涼，切成2片。

餡料

將奶油乳酪和優格攪拌滑順；用滾水把吉利丁粉和糖拌至溶解，加入檸檬汁拌勻攪拌均勻。

組合

1. 將一片胡蘿蔔蛋糕放在糕模內，倒進餡料，再放另一片胡蘿蔔蛋糕，放入冰箱放2小時。
2. 杏仁糕分別與橙色或色綠食用色素搓勻，做成胡蘿蔔狀，取出蛋糕，撒上胡蘿蔔絲和放上胡蘿蔔杏仁糕裝飾。

Carrot cake

1. Sieve all the ingredients of flour and powder. Set aside.
2. Beat butter and sugar until it is fluffy and white. Add egg. Blend thoroughly. Add sieved flours. Add the remainder of cake ingredients. Combine.
3. Pour batter in a cake mould lined with baking paper. Bake at 180°C for 20 minutes. Take out the cake. Let cool. Cut into 2 slices.

Filling

Blend cream cheese and yogurt until smooth. Dissolve gelatin powder and sugar in boiling water. Add lemon juice. Blend until smooth.

Assembly

1. Put a carrot cake slice in a cake mould. Pour in Filling. Put another carrot cake slice. Refrigerate for 2 hours.
2. Knead almond paste with orange and green colourings respectively, forming to a carrot. Take out the cake. Sprinkle with shredded carrot. Garnish with almond paste carrot.

Q：杏仁糕是甚麼？
What is almond paste?

A：杏仁糕是外國蛋糕常用的裝飾，可塑性很高，主要是白色，與食用色素混合而造型，點綴蛋糕。
It is usually used to garnish cakes. It is white in colour and you may knead it with colourings in any shape you want.

Q：這款起士蛋糕的質感輕盈細緻，吃時像風似的，秘密在哪裏？

What is the secret of making this light and silky cheesecake?

A：因為它只用蛋白和玉米粉做主料。蛋白的特質能令蛋糕膨脹和鬆軟，加上用了沒有筋性的玉米粉，口感會更軟泡泡，再添加了起士使到其質感有獨特風味兼變得更柔軟細緻，孔隙很小。

The main ingredients are egg white and cornstarch. Egg white peak add volume and lightness to the consistency of cake. As cornstarch does not contain gluten, it is as soft as delicate foam.

Japanese Style Lemon Cheesecake
檸檬日式起士蛋糕

做法 Directions

奶油乳酪{ 250克 }	(250g) cream cheese
蛋黃{ 4個 }	(4) egg yolks
玉米粉{ 80克 }	(80g) cornstarch
動物性鮮奶油{ 100克 }	(100g) whipping cream
青檸檬{ 半個 }	(½) lime
蛋白{ 4個 }	(4) egg whites
糖{ 80克 }	(80g) sugar

做法 Directions

餡料

1. 將青檸檬刨皮、榨汁，皮和汁分別留用。
2. 將奶油乳酪、蛋黃和動物性鮮奶油攪拌滑順，再拌入玉米粉、青檸檬及汁攪拌滑順。
3. 將蛋白打至濃稠，加入糖繼續打至堅挺細滑，慢慢拌入奶油乳酪鮮奶油混合物攪拌至均勻。
4. 將餡料倒入糕模內，放入烤盤，再倒入熱水，放入已預熱的烤箱中。
5. 先用上火230℃烤約8分鐘，直至金黃，關掉上火，只開下火150℃，繼續烤約20分鐘即可。

Filling

1. Peel and squeeze lime. Keep zest and juice.
2. Blend cream cheese, egg yolk and whipping cream until smooth. Add cornstarch, lime zest and juice. Blend until smooth.
3. Beat egg white until thick. Add sugar. Continue to beat until stiff and smooth. Gradually fold in cream cheese mixture. Combine thoroughly.
4. Pour Filling into a cake mould. Put it in a baking tray filled with hot water. Place it in a preheated oven.
5. Bake with surface temperature at 230°C for 8 minutes until the surface turns to golden. Turn off the heat. Only turn it on with the bottom temperature at 150°C. Continue to bake for 20 minutes.

Q：斑蘭葉汁是甚麼？
What is pandan juice?

A：斑蘭葉是新加坡和馬來西亞的植物，有獨特幽香，可買回來用攪拌機打碎攪汁，也可買入現成斑蘭葉香汁，不過我覺得新鮮的斑蘭葉汁有一陣清香草味，色澤則略淺。
It is a kind of plant from Singapore and Malaysia. It has unique elegant fragrant. Blend it in a blender or buy ready-made juice. Fresh pandan juice has strong grass aroma and light in colour.

難度●●●高

Japanese Style Pandan Cheesecake
斑蘭日式起士蛋糕

材料 Ingredients

奶油乳酪{250克}	(250g) cream cheese
蛋黃{5個}	(5) egg yolks
玉米粉{80克}	(80g) cornstarch
動物性鮮奶油{80克}	(80g) whipping cream
斑蘭葉汁{20克}	(20g) pandan juice
蛋白{5個}	(5) egg whites
糖{100克}	(100g) sugar

做法 Directions

餡料

1. 將奶油乳酪、蛋黃和動物性鮮奶油攪拌滑順，再拌入玉米粉和斑蘭葉汁攪拌滑順。
2. 將蛋白打至濃稠，加入糖繼續打至硬性發泡，慢慢拌入奶油乳酪混合物攪拌直至均勻。
3. 將餡料倒入糕模內，放入烤盤，再倒入熱水，放入已預熱的烤箱中。
4. 先用上火230℃烤約8分鐘，直至金黃，關掉上火，只開下火150℃，續烤約20分鐘即可。

Filling

1. Blend cream cheese, egg yolk and whipping cream until smooth. Fold in cornstarch and pandan juice. Blend until smooth.
2. Beat egg white until thick. Add sugar. Continue to beat until stiff and smooth. Gradually fold in cream cheese mixture. Combine.
3. Pour Filling into a cake mould. Put in a baking tray filled with hot water. Place in a preheated oven.
4. Bake with surface temperature at 230°C for about 8 minutes until golden. Turn off. Only turn it on with the bottom temperature at 150°C. Continue to bake for about 20 minutes.

Q：為什麼要把起士蛋糕放在裝有熱水的烤盤中烘烤，想得到甚麼樣的效果？
Why does cheesecake sit in a tray of hot water in the oven?

A：把起士蛋糕放在裝有熱水的烤盤中烘烤，目的是保持起士蛋糕內裏濕潤的質感。
It is to keep it moist.

難度●●●高

Deluxe Mango Cheesecake
特濃芒果味起士蛋糕

材料 Ingredients

餅底
消化餅底{ 1份 }
（請參閱第122頁）

餡料
奶油乳酪{ 500克 }
糖{ 20克，奶油乳酪用 }
麵粉{ 20克 }
蛋黃{ 5個 }
蛋白{ 5個 }
糖{ 80克，蛋白用 }
冷凍芒果泥{ 100克 }
奶油溶液{ 50克 }

裝飾
芒果粒{ 適量 }

Base
(1) digestive biscuit base
(Refer to p.122)

Filling
(500g) cream cheese
(20g) sugar, for cream cheese
(20g) flour
(5) egg yolks
(5) egg whites
(80g) sugar, for egg white
(100g) frozen mango puree
(50g) butter, melted

Decoration
Mango cubes

做法 Directions

餡料
1. 將奶油乳酪、糖和蛋黃攪拌滑順，再加入麵粉攪拌均勻。
2. 將蛋白打發，再放入糖繼續打至硬性發泡。
3. 將糖蛋白和蛋黃麵粉漿混合，輕輕攪拌滑順，再加入芒果果泥和奶油溶液攪拌滑順。

組合
1. 將消化餅底放入模中，輕輕按實，倒入餡料，放入烤盤，再倒入熱水，用160℃ 烤45分鐘。
2. 取出起士蛋糕，放涼，表面放芒果粒裝飾。

Filling
1. Blend cream cheese, sugar and egg yolk until smooth. Combine with flour.
2. Beat egg white. Add sugar. Continue to beat until stiff.
3. Combine meringue with cream cheese batter. Gently blend. Add mango puree and melted butter. Blend until smooth.

Assembly
1. Put digestive biscuit base in a mould. Press gently with fingertips. Pour in Filling. Place in a baking tray filled with hot water. Bake at 160°C for 45 minutes.
2. Take out the cheesecake. Let cool. Garish with mango cubes.

Q：為何在起士漿添加麵粉成份？
Why do you add flour?

A：因為這款蛋糕內含水果成份，容易出水，使得蛋糕含水份變高了，可以利用少許麵粉有助防止蛋糕過於濕潤，變得水汪汪。
It is to absorb the excessive juice discharged from fruits.

難度●●●高

Rich Blueberry Cheesecake

特濃藍莓起士蛋糕

材料 Ingredients

餅底

消化餅底{1份}
(請參閱第122頁)

餡料

奶油乳酪{500克}
雞蛋{5個}
麵粉{10克}
奶油溶液{50克}
藍莓醬{50克}

裝飾

罐裝藍莓醬{適量}
新鮮藍莓{適量}
哈密瓜球{2粒}
巧克力條{2條}

Base

(1) digestive biscuit base
(Refer to p.122)

Filling

(500g) cream cheese
(5) eggs
(10g) flour
(50g) butter, melted
(50g) blueberry jam

Decoration

Canned blueberry jam
Fresh blueberries
(2) cantaloupe balls
(2) chocolate sticks

做法 Directions

餡料

1. 將奶油乳酪、糖和蛋黃攪拌滑順，再加入麵粉拌勻。
2. 將蛋白打發，再放入糖繼續打至硬性發泡。
3. 將糖蛋白和蛋黃麵粉漿混合，輕輕攪拌滑順，再加入藍莓醬和奶油溶液攪拌滑順。

派底

1. 將消化餅底放入模中，輕輕按實，倒入餡料，放入烤盤，再倒入熱水，用170℃烤1小時。
2. 取出起士蛋糕，放涼，餅面放藍莓漿粒裝飾。

Filling

1. Blend cream cheese, sugar and egg yolk until smooth. Add flour. Combine.
2. Beat egg white. Add sugar. Continue to beat until stiff.
3. Combine meringue with cream cheese batter. Gently blend. Add blueberry jam and melted butter. Blend until smooth.

Assembly

1. Put digestive biscuit base in a mould. Press gently with fingertips. Pour in Filling. Place in a baking tray filled with hot water. Bake at 170°C for 1 hour.
2. Take out the cheesecake. Let cool. Garnish with blueberries.

材料 Ingredients

餅底

曲奇餅底{1份}
（請參閱第122頁）

餡料

奶油乳酪{500克}
糖{20克，奶油乳酪用}
蛋黃{5個}
麵粉{20克}
蛋白{5個}
糖{60克，蛋白用}
黑醋栗汁{100克}
黑醋栗乾{50克}

裝飾

葡萄乾{適量}
鮮果{適量}
巧克力裝飾
已打發植物性鮮奶油{50克}

Base

(1) cookie base
(Refer to p.122)

Filling

(500g) cream cheese
(20g) sugar, for cream cheese
(5) egg yolks
(20g) flour
(5) egg whites
(60g) sugar, for egg white
(100g) black currant juice
(50g) dried black currants

Decoration

Grapes
Fresh fruits
Decorative chocolate
(50g) non-dairy cream, whipped

難度●●●高

Black Currant Cheesecake

黑醋栗起士蛋糕

做法 Directions

餡料

1. 將奶油乳酪、糖和蛋黃攪拌滑順，拌入麵粉攪拌均勻。
2. 將蛋白打發，加入糖打至硬性發泡，慢慢拌入奶油乳酪糊，攪拌滑順。
3. 最後加入黑醋栗汁和黑醋栗乾，攪拌均勻。

組合

1. 將曲奇餅底放入模中，加入餡料，放入烤盤，再倒入熱水，用160℃烤1小時即可。
2. 取出起士蛋糕待涼，用植物性鮮奶油、葡萄乾和鮮果裝飾。

Filling

1. Blend cream cheese, sugar and egg yolk until smooth. Add flour. Combine.
2. Beat egg white. Add sugar. Continue to beat until stiff. Gradually fold in cream cheese mixture. Blend until smooth.
3. Add black currant juice and dried black currants. Combine.

Assembly

1. Put cookie curst in a mold. Add Filling. Put in a baking tray filled with hot water. Bake at 160°C for 1 hour.
2. Take out the cheesecake. Let cool. Garnish with non-dairy cream, grapes and fresh fruits.

Q：為甚麼這個蛋糕的顏色有點怪怪？
How come the colour of the cake looks weird?

A：藍莓、黑醋栗的天然色是深紫，然而當烘焙後它的色素會起了變化，所以出現黑黑藍藍的色澤，這是正常的。
The natural colour of blueberry and black currant is deep purple. After baking, it would change to patchy black and blue colours. It is pretty normal.

Q：為何要加入橙皮於蛋糕內？
Why do you use orange zest?

A：凡是橘子類的水果，它的表皮含橘油成份，經加熱後會散發出甘甜香味，可誘發起士的乳香味道，增加風味。
Orange oil in the rind, after baking, gives sweet aroma. It enhances the creamy taste of cheese.

難度●●●高

Chocolate Chips Cheesecake

巧克力碎粒起士蛋糕

材料 Ingredients

餅底

曲奇餅底{1份}
（請參閱第122頁）

餡料

奶油乳酪{400克}
糖{100克，奶油乳酪用}
高筋麵粉{20克}
雞蛋{2個}
橙汁{60克}
橙皮{半個}
動物性鮮奶油{100克}
巧克力碎片{20克}

裝飾

朱古力粒{適量}
糖霜{適量}
香橙{1個}

Base

(1) cookie base
(Refer to p.122)

Filling

(400g) cream cheese
(100g) sugar, for cream cheese
(20g) bread flour
(2) eggs
(60g) orange juice
(½) orange zest
(100g) whipping cream
(20g) chocolate chips

Decoration

Chocolate drops
Icing sugar
(1) orange

做法 Directions

1. 將奶油乳酪、糖和高筋麵粉攪拌滑順，加入雞蛋和橙汁攪勻，然後再把其餘材料攪拌均勻。
2. 將曲奇餅底放入糕模中，輕輕壓實，倒入餡料，放入烤盤，再倒入熱水。
3. 放入已預熱的入烤箱，用170℃烤1小時即可。
4. 取出起士蛋糕，放涼後，在蛋糕表面裝飾巧克力粒，灑上糖粉，以香橙裝飾。

1. Blend cream cheese, sugar and bread flour until smooth. Add eggs and orange juice. Thoroughly blend. Add the remainder of ingredients. Blend.
2. Put cookie curst in a mold. Press gently with fingertips. Add Filling. Put in a baking tray filled with hot water.
3. Bake at 170°C for 1 hour.
4. Take out the cheesecake. Let cool. Garish with chocolate drops and orange. Sprinkle over top with icing sugar.

材料 Ingredients

餅底

消化餅底{ 1份 }
（請參閱第122頁）

餡料

奶油乳酪{ 500克 }
糖{ 120克 }
麵粉{ 20克 }
蘭姆酒{ 30克 }
檸檬皮{ 半個 }
酸奶油{ 100克 }
巧克力{ 100克 }
動物性鮮奶油
{ 100克，巧克力用 }
雞蛋{ 6個 }

裝飾

巧克力{ 100克 }
動物性鮮奶油{ 50克 }

Base

(1) digestive biscuit base
(Refer to p.122)

Filling

(500g) cream cheese
(120g) sugar
(20g) flour
(30g) rum
(½) lemon zest
(100g) sour cream
(100g) chocolate
(100g) whipping cream, for chocolate
(6) eggs

Decoration

(100g) chocolate
(50g) whipping cream

難度●●●高

Australian Chocolate Cheesecake

澳洲巧克力起士蛋糕

做法 Directions

餡料

1. 將奶油乳酪、糖、麵粉、蘭姆酒、檸檬皮和酸奶油一同打至滑順。
2. 動物性鮮奶油和巧克力一起煮至溶化，拌入奶油乳酪混合物內。
3. 然後將雞蛋逐個加入攪拌滑順至融合。

組合

1. 在糕模裏放入消化餅底，輕輕按實，再倒進餡料。
2. 放入已預熱的烤箱，用170℃烤1小時，取出起士蛋糕放涼。
3. 把巧克力和動物性鮮奶油隔水加熱，待冷卻後，淋在起士蛋糕上，用白巧克力劃線裝飾，即可享用。

Filling

1. Blend cream cheese, sugar, flour, rum, lemon zest and sour cream altogether until smooth.
2. Cook and melt whipping cream and chocolate. Add to cream cheese mixture.
3. Add one egg at a time. Blend until incorporated.

Assembly

1. Put digestive biscuit base in a cake mould. Press gently with fingertips. Pour in Filling.
2. Bake in a preheated oven at 170°C for 1 hour. Take out. Let cool.
3. In a bowl, melt chocolate and whipping cream in hot water. Let cool. Drizzle over top of the cake. Garish with white chocolate. Serve.

Q：如何使巧克力漿能平均地分佈在蛋糕？
How can I spread melted chocolate over top evenly?

A：巧克力淋漿的質感要夠流動，並要一次性淋在蛋糕表面，不能用刮刀抹平，放置在一旁待其凝固。
It must be fluidly. Drizzle all at a time. Do not use scarper. Set aside until it is firm.

Q：這款起士蛋糕愛的味道很濃郁，但卻有熱帶風味，是否加入許多果酸有關？
The flavour of this cheesecake is strong and tropical. Is it because you add fruit acid?

A：是啊！因為它含起士很重，為了平衡起士的香味，保持質感不那麼結實，所以增加了酸奶油和天然果酸如橙汁、青檸汁，亦加重糖份令它比較輕軟。
You are right! To balance out the heavy cheese flavour, I add sour cream and natural fruit acids by using orange and lime juice. I also add sugar in order to make it light and tender.

難度●●●高

Mexican Cheesecake
墨西哥起士蛋糕

材料 Ingredients

餅底

曲奇餅底{1個}
（請參閱第122頁）

餡料1

奶油乳酪{500克}
糖{120克}
酸奶油{30克}
動物性鮮奶油{30克}
蛋黃{3個}
橙汁{30克}
青檸汁{20克}
檸檬皮{半個}

餡料2

雞蛋{3個}

裝飾

檸檬皮
巧克力球{1個}
薄荷葉{數棵}

Base

(1) cookie base
(Refer to p.122)

Filling 1

(500g) cream cheese
(120g) sugar
(30g) sour cream
(30g) whipping cream
(3) egg yolks
(30g) orange juice
(20g) lime juice
(1/2) lime or lemon(?)zest

Filling 2

(3) eggs

Decoration

Lemon zest
(1) chocolate ball
Mint leaves

做法 Directions

餡料

1. 將奶油乳酪、糖、酸奶油、動物性鮮奶油、蛋黃、橙汁、青檸汁和檸檬皮一同打至滑順。
2. 將雞蛋逐個加入奶油乳酪混合物內攪至融合。
3. 將曲奇餅底放入模中，輕輕按實。
4. 倒進奶油乳酪餡料，放入烤盤，再倒入熱水，放入已預熱的烤箱。
5. 用180℃烤1小時即可，取出放涼，用檸檬皮、巧克力球和薄荷葉裝飾。

Filling

1. Blend cream cheese, sugar, sour cream, whipping cream, egg yolk, orange juice, lime juice and lemon zest altogether until smooth.
2. Add one egg at a time to cream cheese mixture. Blend until incorporated.
3. Put cookie curst in a mould. Press gently with fingertips.
4. Pour in cream cheese Filling. Place in a baking tray filled with hot water.
5. Bake in at 180°C for 1 hour. Take out. Let cool. Garnish with lemon zest, chocolate ball and mint leaves.

材料 Ingredients

餅底

消化餅底{1份}
（請參閱第122頁）

餡料

奶油乳酪{500克}
酸奶油{80克}
糖{100克}
麵粉{20克}
雞蛋{6個}
奶油溶液{100克}

裝飾

酸奶油{60克}
蛋白{20克}
糖{10克}

Base

(1) digestive biscuit base (Refer to p.122)

Filling

(500g) cream cheese
(80g) sour cream
(100g) sugar
(20g) flour
(6) eggs
(100g) butter, melted

Decoration

(60g) sour cream
(20g) egg white
(10g) sugar

難度●●●高

Deluxe New York Cheesecake

特濃紐約起士蛋糕

做法 Directions

裝飾

所有材料攪拌均勻。

餡料

1. 把奶油乳酪、酸奶油、糖和麵粉一同攪拌滑順。
2. 將雞蛋逐個打入並攪透，最後拌入奶油溶液。

組合

1. 把消化餅底按入糕模內，輕輕壓實，倒入餡料，放入烤盤，再倒入熱水。
2. 放入已預熱的烤箱，用170℃烘烤1小時。
3. 取出起士蛋糕，抹上裝飾酸奶油，入烤箱烤3分鐘200℃上火，下火不要即可。

Decoration

Combine sour cream with egg white and sugar for Decoration. Set aside.

Filling

1. Blend cream cheese, sour cream, sugar and flour until smooth.
2. Add one egg at a time. Blend until incorporated. Add melted butter.

Assembly

1. Put digestive biscuit base in a cake mould. Press gently with fingertips. Pour in Filling. Place in a baking tray filled with hot water.
2. Bake in a preheated oven at 170°C for 1 hour.
3. Take out the cheesecake. Spread top with sour cream. Done. Transfer to the oven and bake for 3 minutes under top fire at 200°C without bottom fire.

Q：為何要將起士蛋糕隔熱水烘烤，不能用冷水？
Why do you use hot water instead of cold water?

A：把生的起士蛋糕放在熱水上再送入烤箱，以確保烤箱溫度和蛋糕溫度差距不大，不影響到烘烤蛋糕受熱情況。
The purpose is to narrow the difference in temperature between the oven and cake, and keep the heat in control.

Q：咖啡鮮奶油如何做？
How to make coffee cream?

A：只要把濃咖啡溶液與已打發的植物性鮮奶油混合即可，一般製作的狀況是20-30克咖啡溶液，可配合100-150克已打發的鮮奶油。
Mix 20-30g strong coffee liquid and 100-150g whipped non-dairy cream.

難度●●●高

Charcoal Roasted Coffee Cheesecake
炭燒咖啡起士蛋糕

材料 Ingredients

餅底

消化餅底{1份}
（請參閱第122頁）

餡料

奶油乳酪{500克}
糖{100克}
玉米粉{20克}
炭燒咖啡{15克}
滾水{10克，炭燒咖啡用}
雞蛋{5個}
奶油溶液{80克}

Base

(1) digestive biscuit base (Refer to p.122)

Filling

(500g) cream cheese
(100g) sugar
(20g) cornstarch
(15g) charcoal roasted coffee
(10g) boiling water, for coffee
(5) eggs
(80g) butter, melted

做法 Directions

餡料

1. 把炭燒咖啡和滾水調勻。
2. 將奶油乳酪、糖、玉米粉和炭燒咖啡水攪拌均勻。
3. 將雞蛋依序加入，直至完全攪透，最後加入奶油溶液，攪拌滑順。

組合

1. 在糕模裏放進消化餅底，輕輕按實，倒入奶油乳酪餡後，放入烤盤，再倒入熱水，放進已預熱的烤箱。
2. 用170℃的烤箱烤1小時，取出，放涼。
3. 可擠上咖啡鮮奶油或用鮮果和巧克力條裝飾。

Filling

1. Combine charcoal roasted coffee with boiling water.
2. Combine cream cheese with sugar, cornstarch and coffee.
3. Add one egg at a time. Blend until incorporated. Add melted butter. Blend until smooth.

Assembly

1. Put digestive biscuit base in a cake mould. Press gently with fingertips. Pour in Filling. Place in a baking tray filled with hot water.
2. Bake at 170°C for 1 hour. Take out. let cool.
3. Pipe coffee cream. Or garnish with fresh fruits and chocolate stick.

Q：什麼時候可以在蛋糕上劃花紋？
How can I make pattern on the cake?

A：在蛋糕漿上可以劃花紋，只要它是不同顏色，而蛋糕漿本身的密度高即可。
Just mix batter of different colours and make sure the consistency is thick enough.

難度●●●高

American Cheesecake
美國濃味起士蛋糕

材料 Ingredients

餅底
消化餅底{1份}
（請參閱第122頁）

餡料
奶油乳酪{500克}
糖{150克}
雞蛋{6個}
奶油溶液{100克}

裝飾
黑巧克力{20克}

Base
(1) digestive biscuit base
(Refer to p.122)

Filling
(500g) cream cheese
(150g) sugar
(6) eggs
(100g) butter, melted

Decoration
(20g) dark chocolate

做法 Directions

餅面
把巧克力以熱水隔水加熱至溶解。

餡料
1. 將奶油乳酪在室溫下放軟，與糖混合攪拌至滑順。
2. 逐個加入雞蛋攪拌至完成。
3. 最後加入蛋和奶油溶液，攪拌滑順。

組合
1. 在糕模裏放進消化餅底，輕輕按實，再倒入起士餡，並把巧克力溶液倒入，以竹籤劃花後，放入烤盤，再倒入熱水，再放進已預熱的烤箱。
2. 用170℃的烤箱烤1小時，取出，放涼。

Decoration
In a bowl, melt chocolate in hot water.

Filling
1. Leave cream cheese at room temperature until softened. Blend with sugar until smooth.
2. Add one egg at one time, blending thoroughly.
3. Add egg and melted. Blend until smooth.

Assembly
1. Put digestive biscuit base in a cake mould. Press gently with fingertips. Pour in cheese Filling. Pour in melted chocolate. Briefly stir to make patterns with a bamboo skewer. Put in a baking sheet filled with hot water. Place in a preheated oven.
2. Bake at 170°C for 1 hour. Take out. Let cool.

材料 Ingredients

巧克力蛋糕

黑巧克力{80克}
奶油{30克}
雞蛋{3個}
糖{45克}
麵粉{20克}
香草精{少許}

起士餡料

奶油乳酪{125克}
糖{20克}
雞蛋{1個}
動物性鮮奶油{100克}
奶油溶液{10克

Chocolate cake

(80g) dark chocolate
(30g) butter
(3) eggs
(45g) sugar
(20g) flour
Vanilla extract

Cheese Filling

(125g) cream cheese
(20g) sugar
(1) egg
(100g) whipping cream
(10g) butter, melted

難度●●●高

Double Layer Cheesecake

雙層起士蛋糕

做法 Directions

巧克力蛋糕

1. 巧克力與奶油一起隔水加熱。
2. 雞蛋和糖一同攪拌至濃稠，慢慢拌入麵粉和香草精。

起士餡料

將奶油乳酪和糖攪拌滑順，逐個雞蛋加入攪拌滑順，拌入動物性鮮奶油和奶油溶液，攪拌滑順成起士餡。

組合

1. 把巧克力蛋糕倒入糕模內，放入已預熱的烤箱。
2. 用180℃的烤箱烤約15分鐘，出烤箱，倒入起士餡，回烤箱繼續烤20分鐘。

Chocolate cake

1. Melt chocolate and butter together.
2. Blend egg and sugar until thick. Gradually fold in flour and vanilla extract.

Cheese Filling

Blend cream cheese and sugar until smooth. Add one egg at a time, blending until smooth. Fold in whipping cream and melted butter. Blend until smooth and it forms to cheese Filling.

Assembly

1. Pour chocolate batter into cake mould. Put in a preheated oven.
2. Bake at 180°C for 15 minutes. Remove from the oven. Pour in cheese Filling. Bake for another 20 minutes.

Q：為何要把雞蛋逐個加入起士漿內？
Why do you add one egg to cheese mixture at a time?

A：只有這樣做雞蛋才能完全融入起士混合物，直到烘烤時蛋漿能均勻地散佈蛋糕內，使其質感鬆軟有雞蛋香。
In doing so, you can make sure that eggs incorporate into the cheese mixture. The baked cheesecake is soft with full egg aroma.

起士蛋糕房訓練室
Cheesecake Baking Workshop

一份奶油乳酪，加點雞蛋攪攪拌拌，再倒入已溶化的吉利丁粉輕輕拌入，噢！就快做好了。快些取餅模，倒進混合物，放冰箱待凝固，很快便有得吃啦！喂！不准偷吃……看似簡單的程序，但不認識工具和基本技巧，如何能做出一個美味的起士蛋糕？所以你應該了解這些基本常識，才能按步就班，量身訂做你的稱心蛋糕啊！

Add eggs to cream cheese. Blend. Gently mix with gelatin solution. Oh, will get it done soon! Get a cake mould. Pour in the mixture. Chill. We may serve it very soon! Hey, don't sneak eating… They sound like simple steps. But how can you make a yummy cake without the knowledge of tools and basic skills? It is necessary to learn the basics and make a lovely cake step by step!

必備工具和材料 Tools and Ingredients

1. 烘焙餅架(放涼架)：用來承托剛出爐的糕餅，讓空氣可以在蛋糕底部流通，糕餅涼後就會乾爽。

   Wire rack (Cooling Rack):This is designed to hold a cake or other baked item above the surface of a kitchen counter so air can circulate under the cake and allow it to dry while cooling. Otherwise it may become soggy because the moisture cannot escape from the bottom.

2-3. 餅乾模/牙鈒(大/中)：把桿平的麵糰切割造型。

   Cookie Cutter (Large/Medium): This is a tool to cut out pastry in particular shape.

4. 舒芙蕾烤杯：圓形陶瓷的耐熱烤杯，可放進烤箱烘焙，也有Medline紙製舒芙蕾杯。

   Soufflé Cup: This is made with porcelain, for baking Soufflé. There are also Medline Paper Soufflé Cup.

5. 奶油蛋糕長模：長方形高身烘烤餅模，可供烘焙磅蛋糕或麵包。

   Loaf Tin: Straight sided rectangular cake tin for baking pound cake or bread.

6. 6吋圓形蛋糕模：直徑6吋的圓形蛋糕模具是香港現今流行的尺寸。

   6-in Round Cake Tin (Baking Mould): This is a popular size in Hong Kong.

7. 6吋無底圓形蛋糕模：可供製作多種糕餅，用烘焙紙墊底，脱模方便。

   6-in Bottomless Cake Tin (Mousse Ring): It is for baking of various simple pastries such as base cake or pastry shell, directly on baking sheet.

8. 三角形無底圓形蛋糕模：用以製作三角形的糕餅。

   Triangular Bottomless Cake Tin: It is for making triangular shape cake.

9. 擠花袋：可以裝入鮮奶油、麵糊等，擠在蛋糕或烤盤上。有尼龍質和膠質的。

   Piping Bag: Use with a varity of nozzles, it is for piping icing, cream, potatoes and pastry. Made from Nylon or plastic.

10. 粉篩：用來篩粉至細滑，隔去雜質的器具。

    Sifter: It is for sifting different kinds of flour and powder.

11. 電子秤：用來秤量大量食材，如麵粉。

    Electric Scale: Use for measuring weight and large scale ingredients, such as flour.

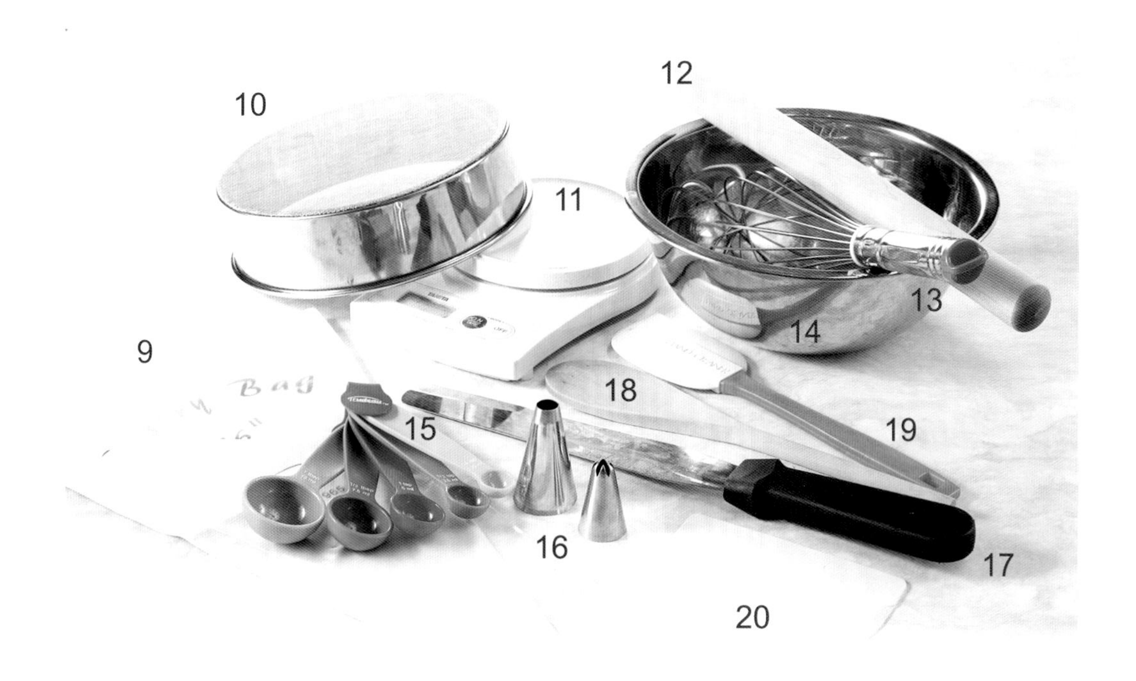

12. 桿麵棍：平滑、質感重、無手柄，用來桿薄麵糰。

    Rolling Pin: A good rolling pin should be plain and smooth. Made by heavy hardwood, with or without handle. Some rolling pin handles are fixed to a central rod in the roller.

13. 攪拌器(手提打蛋器)：打發混合材料的器具。

    Whisk (Handheld Beater): It is a tool for whisking and mixing.

14. 不鏽鋼盆：清洗方便，是用作打發、攪拌和混合材料的容器。

    Stainless Mixing Bowl: It is container for whisking, blending and combining ingredients. Easy to clean.

15. 量匙：用來量少量材料，如糖、油等。

    Measuring Spoons:Use for measuring small scale ingredients.

16. 金屬擠花嘴：有不同尺寸、花樣，能擠出不同花紋。

    Metal Nozzles (Piping Tube): Nozzles are for piping different kinds of shape and pattern of batter or cream. Available in different sizes.

17. 抹刀：用來塗抹鮮奶油或起士於蛋糕上。

    Palette Knife: It is a non-sharp knife to spread cream or cheese on the cake.

18. 木匙：用於攪拌、混合物料。因木匙不容易傳熱，用來煮高溫食物較為適合。

    Wooden Spoon: It is ideal for stirring and mixing ingredients. Wood is not easily to transit heat by handle that it is suitable for long cooking.

19. 膠刮刀：可用作攪拌麵糊、混合材料。

    Plastic Spatula: It is for folding ingredients and batter and scraping ingredients out of mixing bowl.

20. 膠抹刀/膠切刀：用於切割麵糰 ，混合以及抹平麵糊，也可將麵糊抹出。

    Plastic Palette: It is for cutting dough, folding ingredients and scraping batter and ingredients out of mixing bowl.

缺不了的原料與處理 Basic ingredients and process

攪打鮮奶油(甜、淡)、溶化吉利丁(粉狀、片狀)、巧克力(含巧克力油成份)

how to whip cream (non-dairy and whipping cream), how to dissolve gelatin (powder and sheet), chocolate (that conutter)

動物性鮮奶油和植物性鮮奶油 Whipping cream and non-dairy cream

動物性鮮奶油含35%油脂，味道淡而含富奶香味道，質感細滑，主要用作做餡和放湯為主，不過打發時要小心呀！因為當你看像水一樣稀稀的狀態時，需要繼續打發，待出現濃稠狀就要小心了，因為數秒間鮮奶油會突變成固體，要是仍然繼續攪打，它會因過度打發而出現分離現象，鮮奶油不但成不了形，還會變成粗糙，不能再使用了，需要丟棄，重新再來。

植物性鮮奶油不是真鮮奶油，屬含甜味的裝飾用鮮奶油，容易打發，質感打起後堅挺，不易下塌，對於初學者會比較容易處理，價錢比較實惠，但相對於動物性鮮奶油的質感就沒那麼細滑。

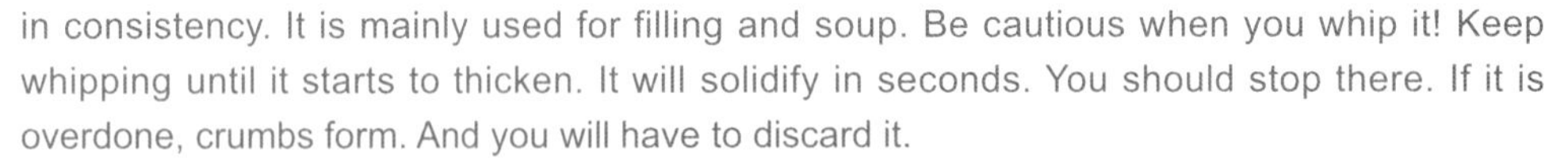

Whipping cream contains 35% fat. The taste is light with full creamy aroma. It is smooth in consistency. It is mainly used for filling and soup. Be cautious when you whip it! Keep whipping until it starts to thicken. It will solidify in seconds. You should stop there. If it is overdone, crumbs form. And you will have to discard it.

Non-dairy cream is not real cream. It is sweetened cream for decoration. It is easy to whip and form to peak. It is stiff and does not collapse easily. For beginners, it is a good option as it is also cheaper than whipping cream. It is not so smooth, though.

吉利丁粉 Gelatin powder

吉利丁粉是由海藻提煉而成，粉末狀而價錢實惠，最適合新手使用。但單把清水與它混合，不易溶解。解決方法是先與少許砂糖混合，再放清水慢慢攪拌，比較容易融和，要是還不能完全融合，就以熱水隔水加熱中幫助溶解。

Gelatin powder is refined from algae. It is powder form and the price is reasonable. I' d suggest beginners to use it. It is hard to dissolve it in water alone. The trick is to mix with sugar and then slowly stir in water. It is easier to incorporate. If not, warm it up in a bowl of hot water.

烘焙巧克力、白巧克力和黑巧克力

Chocolate for baking, white chocolate and dark chocolate

烘焙巧克力是半甜的巧克力，它是用黑巧克力和糖各佔一半製造而成，質感比較硬，味道有點微苦和濃郁，可供烹煮和耐高溫而不易溶化，多用作曲奇和鬆餅。

白巧克力是假巧克力，以糖、牛奶和可可油混合而成，缺少了可可固醇，屬糖果類製品，多用作裝飾，所以沒有巧克力香味，味道比較甜。

黑巧克力是真正巧克力，又稱純巧克力或黑巧克力，依據含可可油成份而決定味道的濃郁度，一般以55%至72%，最適合做烘焙糕餅，要是濃度超過這個幅度就過濃了，苦味也很強，除非你很愛吃苦味巧克力，對於初學者就要用55%最理想。

Chocolate for baking is semi-sweetened chocolate, containing equal portions of dark chocolate and sugar. The taste is a little bit bitter and rich. It is ideal for cooking, such as making cookies and muffins, at high temperature as it does not melt easily.

White chocolate is a chocolate derivative as it consists of sugar, milk and cocoa butter, but not cocoa solids. It is commonly used for decoration. Taste is sweet without the chocolate aroma.

Dark chocolate, aka pure chocolate, is real chocolate. Its richness varies according to the content of cocoa butter, usually between 55% and 72%. It is the best option for baking cakes. Bitterness will be overwhelming if the content of coca butter is over 72%, unless bitter chocolate is your flavour. Beginners are advised to use 55% dark chocolate.

巧克力裝飾 Chocolate decoration

市面上有很多巧克力裝飾，形狀也有很多選擇，有長片和三角片主要做圍邊、葉形、花形、球形、捲曲紋、長管和褶花紋形等就用作蛋糕面裝飾，可按蛋糕的形態選擇所需，簡單方便，容易損壞，價錢比較昂貴。

A variety of chocolate products for decoration are available in the market. Panels and triangles are mainly used to decoration the sides of cake. You may garnish top with leaves, flowers, balls, curls, cigarellos, folds, etc, up to the shape of cake. Handle with care as it is easy to break them. The prices are more expensive.

摺蛋糕模的紙底

How to fold parchment paper and plastic wrap

(一) 鋪烘焙紙 Parchment paper lining

1. 先備一張長方形的奶油紙或烘焙紙，摺起一個三角形。
2. 沿邊緣摺小三角形，記緊摺紙時必須緊貼餅模，不能太鬆，否則倒進蛋糊時會出現缺口，不能成形。
3. 摺到最尾位時，把紙尾端扭實，再把它收好。

1. Fold a rectangular parchment paper into a triangle.
2. Along with the edge, fold it up with little triangles tightly. If it is loose, batter will spill over and leak into gaps and cake cannot form in a desired shape.
3. When it comes to the last fold, twist and seal tightly.

(二) 墊保鮮膜 Plastic wrap lining

1. 在糕模先墊上一張保鮮膜。
2. 然後把收口扭好，記得倒漿前必須放在烤盤，保持餡料不會流出來。

1. Line a cake mould with a piece of plastic wrap.
2. Twist the opening end. Remember to line it before baking.

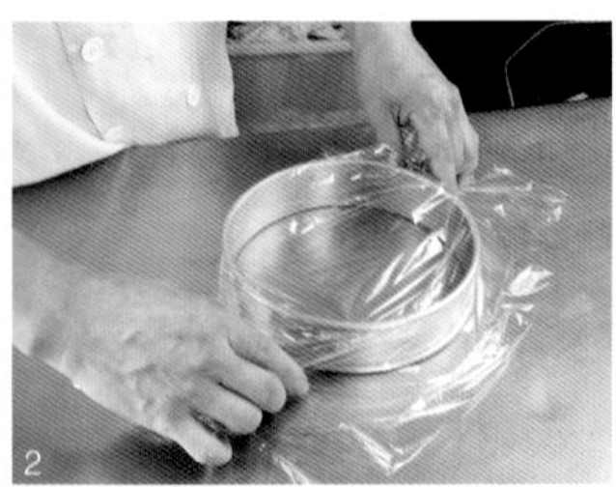

新手必懂，餅底齊齊做

Base for begginers

清蛋糕或海綿蛋糕 Base cake

餡料

雞蛋{2個}
糖{40克}
麵粉{40克}
奶油溶液{15克}

Ingredients

(2) eggs
(40g) sugar
(40g) flour
(15g) melted butter

做法

1. 雞蛋和糖同放在器皿上，然後打至濃稠，以把它拉起來時能在蛋糊的表面寫字為準。
2. 把麵粉過篩，慢慢以膠刮刀拌入麵粉，使空氣隨拌入粉加入蛋糊內，有助蛋糕打發。
3. 最後加入奶油溶液，一旦拌勻後就要迅速倒入已墊紙或刷油撲粉的糕模內，放入已預熱的烤箱，確保烤箱溫度穩定，以180℃烘焙15分鐘直至金黃，用竹籤試插，如沒有蛋糕漿黏貼，表示已經熟透，取出翻轉，待涼之後，可保持表面平坦。

   註：與巧克力蛋糕的做法一樣，只是麵粉要扣掉5克，改加入巧克力粉5克取代。

Directions

1. In a bowl, beat eggs and sugar until thickened and you may make patterns on the surface of the beaten egg.
2. Sieve flour. With a plastic scraper, add flour. More air fold in the beaten egg and help it expand.
3. Add melted butter. Blend and quickly pour in a lined or greased and floured cake mould, Bake in a preheated oven (let the temperature stable) at 180℃ for 15 minutes until golden. Test with a bamboo stick. If no batter sticks to it, it is done. Take out and turn over. Let cool. The surface may keep level.

Note: To make chocolate cake, 5g cocoa powder replaces 5g flour.

糖酥 / 鬆酥 / 派皮底 Sweet pastry / Shortcrust pastry / Pie base

材料

奶油{ 30克 }
糖{ 10克 }
麵粉{ 50克 }
雞蛋{ 10克 }

做法

1. 把奶油和糖拌勻，加入雞蛋搓勻，再與麵粉搓揉成糰。
2. 將麵糰放入糕模，輕輕壓實。
3. 放進已預熱的烤箱，用200℃烘焙15分鐘直至變金黃和酥脆，因為想有酥脆和較乾爽的效果，所以採用高一點的烤箱溫度處理，目的是為了有吃餅乾的口感。

Ingredients

(30g) butter
(10g) sugar
(50g) flour
(10g) egg

Directions

1. Blend butter and sugar. Combine with egg. Knead with flour and it forms to dough.
2. Put dough in a cake mould. Gently press with fingertips.
3. Bake in a preheated oven at 200℃ for 15 minutes until golden and crisp. High oven temperature is to increase the crispness.

1

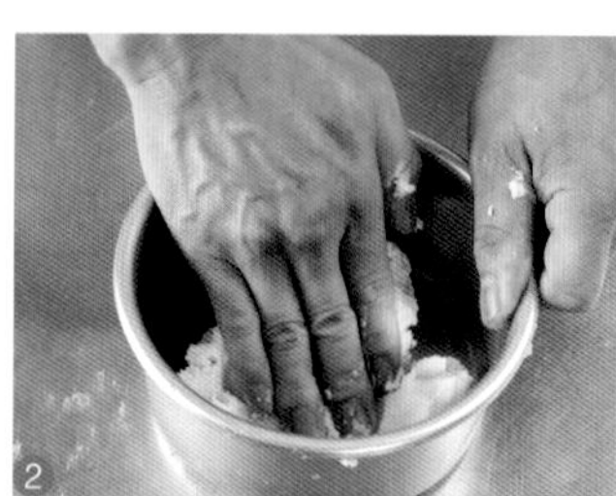
2

3

消化餅 / 茶餅底 Digestive biscuit / Tea biscuit base

材料

麥維他消化餅{ 80克 }
奶油溶液{ 40克 }

做法

1. 將消化餅壓碎，可買現成貨，但我認為它太細碎，所以有時間自己磨碎比較好，選擇也很多。
2. 加入奶油溶液，把所有物料拌勻，應該有點微濕且能成糰為佳。
3. 倒進糕模內，用湯匙輕輕壓實，放冰箱使餅底凝固。

Ingredients

(80g) digestive biscuits
(40g) melted butter

Directions

1. Crush digestive biscuits. I prefer to do it myself. If time does not allow, you may pick dy-made crumbs, though.
2. Add melted butter. Combine all the ingredients. Mix well. It forms to slightly moist dough.
3. Pour in a cake mould. Gently press with a spoon. Refrigerate until firm.

1

2

3

4

5

起士蛋糕的拌合法 How to blend cheesecake ingredients

(一)烤起士蛋糕 Baked cheesecake

材料

奶油乳酪{500克}
雞蛋{6個}
糖{150克}
奶油溶液{100克}

Ingredients

(500g) cream cheese
(6) eggs
(150g) sugar
(100g) melted butter

做法

1. 奶油乳酪與糖拌勻，因為有糖的成份，奶油乳酪比較容易混合融和。
2. 加入雞蛋，用點力攪拌至完全融入，其質如細滑鮮奶油狀。
3. 倒入奶油溶液拌勻，以不見油浮於起士漿為準。
4. 放入已預熱的烤箱，以170℃烤1小時，用慢火烘烤起士蛋糕，目的是保持蛋糕的質感滑如絲綢，不會變粗糙。按烘焙效果的質感和濕度而決定是否需要隔著熱水烘焙。

Directions

1. Blend cream cheese and sugar. With sugar, it is easy to blend cream cheese completely.
2. Add egg. Slightly vigorously blend until incorporated completely. The consistency is smooth and creamy.
3. Pour in melted butter. Blend until no butter floats on the surface.
4. Bake in a preheated oven at 170℃ for 1 hour. Low oven temperature is to keep cheesecake silky and smooth. Baking with a tray of hot water is optional.

（二）冰起士蛋糕 Chilled cheesecake

材料

海綿蛋糕{2片}
（參閱第121頁）
奶油乳酪{125克}
糖{20克}
雞蛋{2個}
糖{20克}
吉利丁粉{12克}
滾水{70克}
植物性鮮奶油
{150克，打發}

做法

1. 放一片海綿蛋糕。
2. 奶油乳酪壓碎，放入少許糖拌勻成鮮奶油狀。
3. 雞蛋與其餘糖在熱水中隔水加熱打發，有助殺死蛋黃中的沙門氏菌。一般情況下，蛋黃的溫度達65℃以上已乎合食物安全的要求，又不會弄熟雞蛋而令質感變粗糙，再與起士混合物融合。
4. 將動物性鮮奶油和植物性鮮奶油混合，倒入少許蛋黃起士醬融和，再放於鮮奶油混合物。
5. 吉利丁粉與滾水拌溶，再與起士混合物拌勻，混合時要快速攪拌。如見吉利丁水傾向凝固現象，可坐熱水變回液態狀。
6. 倒半份起士混合物於已墊海綿蛋糕片的模具，再於另一片海綿蛋糕，輕輕按實，切記蛋糕片要比糕模略小，否則容易把起士混合流瀉。
7. 再倒入剩餘起士混合物，然後輕輕搖勻，平放後置冰箱冷卻凝固。

Ingredients

(2) slices base cake
(Refer to p.121)
(125g) cream cheese
(20g) sugar
(2) eggs
(20g) sugar
(12g) gelatin powder
(70g) boiling water
(150g) non-dairy cream,whipped

Directions

1. Place a base cake slice in a cake mould.
2. Crush cream cheese. Add a small amount of sugar. Blend until creamy.
3. In a bowl, blend egg and the remaining sugar in hot water until smooth. It may kill salmonella in eggs. Generally speaking, 65℃ meets to requirement of food safety. Egg is not cooked and crumbs do not form at this temperature before mixing with cream cheese mixture.
4. Combine whipping cream and non-dairy cream. Add a small amount of cream cheese mixture. Blend until incorporated. Add ? to mixed cream again.
5. Dissolve gelatin powder in boiling water. Combine with cream cheese mixture. Blend fast. If crumbs appear, warm up in hot water.
6. Pour ½ of cream cheese mixture into a cake mould lined with base cake. Place another base cake slice on top. Gently press with fingertips. The cake slice should be slightly smaller than the cake mould. If not, the cream cheese mixture may spill over.
7. Pour in the remaining cream cheese mixture. Gently shake the mould. Place on a table to level the surface. Refrigerate until firm.

1

2

3

4

5

6

7

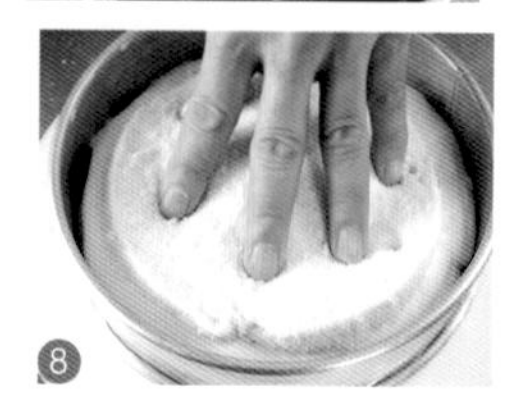
8

9

10

擠花嘴與擠花技巧

Pastry tip and piping basics

糕餅製作好玩之處，就是從零開始，自由發揮空間很大，擁有個人風格，成就感很大，所以如何擅用擠花嘴來裝飾蛋糕，好玩又創意無窮，簡單的數個唧嘴，就有千變萬化的感覺。

The fun of making cakes is that you start from scratch with plenty room of creativity and personal style. It gives you a great sense of achievement. With some simple pastry tips, you may create unlimited decorating ideas!

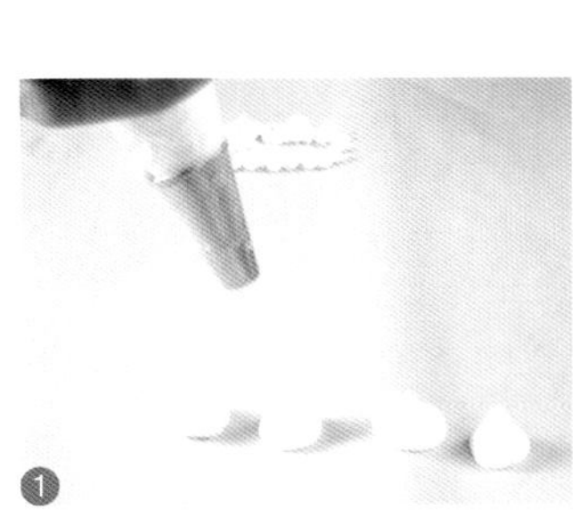

❶ 圓平嘴 Plain round tip

❷

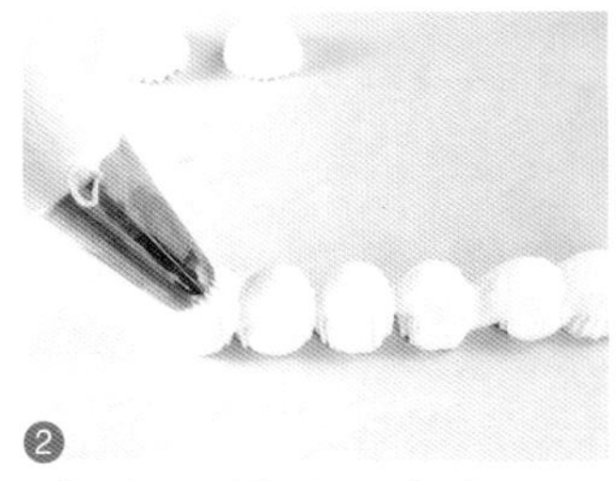

❷

菊花嘴，線紋細密、圓拱形 Chamomile-shaped tip, to make fine lines and arch.

❸ 星形嘴 Star tip

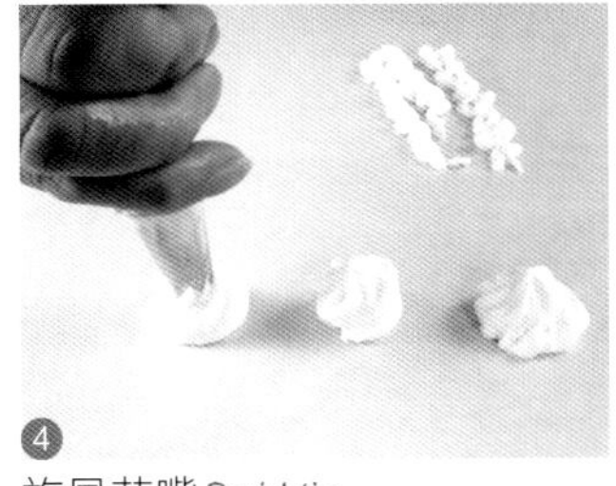

❹ 旋風花嘴 Swirl tip

❺ 葉形嘴 Leaf tip

❻ 花瓣嘴 Petal tip

第一次學做
Cheesecake!
Yummy

作　　者　邱勇靈
發 行 人　程安琪
總 策 劃　程顯灝
執行編輯　譽緻國際美學企業社、盧美娜
主　　編　譽緻國際美學企業社、莊旻嬑
美　　編　譽緻國際美學企業社
封面設計　洪瑞伯

出 版 者　橘子文化事業有限公司
總 代 理　三友圖書有限公司
地　　址　106 台北市安和路2段213號4樓
電　　話　(02) 2377-4155
傳　　真　(02) 2377-4355
E-mail　service@sanyau.com.tw
郵政劃撥　5844889 三友圖書有限公司

總 經 銷　大和書報圖書股份有限公司
地　　址　新北市新莊區五工五路2號
電　　話　(02) 8990-2588
傳　　真　(02) 2299-7900

SAN YAU
http://www.ju-zi.com.tw
三友圖書
友直 友諒 友多聞

初　　版　2014年02月
定　　價　新臺幣 298元
I S B N　978-986-6062-70-4

本書由香港萬里機構出版有限公司
授權在台灣出版發行

國家圖書館出版品預行編目(CIP)

第一次學做Chessecake!Yummy / 邱勇靈作. -- 初版. -- 臺北市 : 橘子文化, 2014.02
面；　公分

ISBN 978-986-6062-70-4(平裝)

1.點心食譜

427.16　　102026186

地址：　　　縣/市　　　　鄉/鎮/市/區　　　　路/街

段　　巷　　弄　　號　　樓

廣　告　回　函
台北郵局登記證
台北廣字第2780號

SANYAU

三友圖書有限公司 收

SANYAU PUBLISHING CO., LTD.

106　台北市安和路2段213號4樓

SANYAU

三友圖書 / 讀者俱樂部

填妥本問卷，並寄回，即可成為三友圖書會員。
我們將優先提供相關優惠活動訊息給您。

粉絲招募
歡迎加入

◦看書 所有出版品應有盡有
◦分享 與作者最直接的交談
◦資訊 好書特惠馬上就知道

旗林文化╳橘子文化╳四塊玉文創
https://www.facebook.com/comehomelife

【黏貼處】對折後寄回，謝謝。

親愛的讀者：

感謝您購買《第一次學做Chessecake!Yummy》一書，為感謝您的支持與愛護，只要填妥本回函，並寄回本社，即可成為三友圖書會員，將定時提供新書資訊及各種優惠給您。

1 您從何處購得本書？
□博客來網路書店 □金石堂網路書店 □誠品網路書店 □其他網路書店
□實體書店______

2 您從何處得知本書？
□廣播媒體 □臉書 □朋友推薦 □博客來網路書店 □金石堂網路書店
□誠品網路書店 □其他網路書店______□實體書店______

3 您購買本書的因素有哪些？(可複選)
□作者 □內容 □圖片 □版面編排 □其他______

4 您覺得本書的封面設計如何？
□非常滿意 □滿意 □普通 □很差 □其他______

5 非常感謝您購買此書，您還對哪些主題有興趣？(可複選)
□中西食譜 □點心烘焙 □飲品類 □瘦身美容 □手作DIY
□養生保健 □兩性關係 □心靈療癒 □小說 □其他______

6 您最常選擇購書的通路是以下哪一個？
□誠品實體書店 □金石堂實體書店 □博客來網路書店 □誠品網路書店
□金石堂網路書店 □PC HOME網路書店 □Costco
□其他網路書店______ □其他實體書店______

7 若本書出版形式為電子書，您的購買意願？
□會購買 □不一定會購買 □視價格考慮是否購買 □不會購買
□其他______

8 您是否有閱讀電子書的習慣？
□有，已習慣看電子書 □偶爾會看 □沒有，不習慣看電子書
□其他______

9 您認為本書尚需改進之處？以及對我們的意見？
____

10 日後若有優惠訊息，您希望我們以何種方式通知您？
□電話 □E-mail □簡訊 □書面宣傳寄送至貴府 □其他______

謝謝您的填寫，
您寶貴的建議是我們進步的動力！

姓名______________ 出生年月日______________
電話______________ E-mail______________
通訊地址______________________________

【黏貼處】對折後寄回，謝謝。